The Internet
in the Middle East

SUNY series in Computer-Mediated Communication
Teresa M. Harrison and Timothy Stephen, Editors

The Internet
in the Middle East

Global Expectations and
Local Imaginations in Kuwait

Deborah L. Wheeler

STATE UNIVERSITY OF NEW YORK PRESS

μL
Published by
State University of New York Press, Albany

© 2006 State University of New York

For information, address State University of New York Press,
90 State Street, Suite 700, Albany, N.Y. 12207

Production by Diane Ganeles
Marketing by Susan M. Petrie

Library of Congress Cataloging-in-Publication Data

Wheeler, Deborah L., 1965–
    The Internet in the Middle East : global expectations and local imaginations
in Kuwait / Deborah L. Wheeler
        p. cm. — (SUNY series in computer-mediated communication)
    Includes bibliographical references and index.
    ISBN 0-7914-6585-3 (hardcover : alk. paper) — ISBN 0-7914-6586-1 (pbk :
    alk. paper)
        1. Internet—Kuwait.   2. Cyberspace—Middle East.   3. Telecommunication—
Social aspects—Arab countries.   I. Title.   II. Series.

TK5105.875.I57W523 2005
300.48'33'09174927—dc22   2004027306

To my mother, Barbara J. McNair,
for everything and more;
and in memory of
Connie Wheeler Reiff (1941–1994),
without whom there is less laughter.

Contents

Acknowledgments

This is a book about the Internet, about Kuwait, and about the Internet in Kuwait. In writing this book over the past seven years, I accumulated a number of debts. Due to the generous support of a number of institutions, none of them are financial. I would like to thank the Council on the Foreign Exchange of Scholars and Kuwait University for their support in the form of a Fulbright Post Doctoral Research Fellowship, 1996-1997. I also thank the American Cultural Center Damascus, the Fulbright Bi-National Commission Morocco, and the British Council, for organizing an international lecture tour in 1997 which gave me an opportunity to share my preliminary research findings on the Internet in Kuwait with a global audience. The Kuwait Foundation for the Advancement of Science and the Kuwait Institute for Scientific Research offered support for my return visit to Kuwait in 1998 as a participant in Kuwait's Conference on Information Superhighway. Three institutions provided fellowships that facilitated the time and financial resources needed to write this book, including the University of Washington's Center for Internet Studies (2000–2001), the Berglund Center for Internet Studies (2001–2002), and the Oxford Internet Institute at the University of Oxford (2003–2005).

Over 100 Kuwaitis gave interviews and shared their opinions about my research questions, and for their support and frankness, I am indebted. In deference to their privacy I will not be able to acknowledge all of them openly. Several Kuwait colleagues went out of their way to make me feel at home in their country. These individuals include the late Dr. Saif Abbas, Khulud al-Fili, Dr. Siham al-Freeh, and her husband, Badr al-Rifai, Dr. Ghada

Hijawi al-Qadummi and her late husband, Hani al-Qad-ummi, Dr. Muhammad al-Rumayhi, Dr. Ahmad al-Bagh-dadi, Dr. Shamlan al-Essa, Dr. Masuma al-Mubarak, Dr. Fahad al-Mekrad, Dr. Shafiq al-Ghabra, Dr. Khaldun al-Nakib, Dr. Hassan al-Johar, Dr. Hussain al-Ansari, and Dr. Yusif al-Ibrahim. I also would like to thank several press organizations for access to their news archives in-cluding the Kuwait News Agency, *al-Anba'*, *al-Siyassa*, and *al-Rai al-Am*.

A number of scholars in the United States and Britain have aided in the research, writing and revising of this manuscript. Two anonymous reviewers at State Univer-sity of New York Press are to be thanked. I also thank col-leagues who read and commented on the manuscript, in-cluding Mary Ann Tétreault, Charles Ess, Kathleen D. Noble, Robert Rook, Raymond Bush, Naomi Sakr, Miriam Lips, Paula Holmes Eber, Nancy Baran Mickle, Brannon M. Wheeler, Susanne Hoeber Rudolph, and Stephen Cole-man. I thank Bjorn Anderson of the University of Michi-gan as well for making the map of Kuwait for chapter 3. I would like to thank Mary Ann Tétreaut and Fawaz al-Sirri for the cover photos.

My spouse, Brannon M. Wheeler, has been a constant source of inspiration and support throughout the re-search and writing of this manuscript and throughout life. I thank our children, Jeffry, Zachary, and Franklin, for putting up with their parents' incessant need for for-eign travel and study. Many of the insights found within these pages are a direct result of seeing the world through their eyes. Of course, any shortcomings in this narrative I own alone.

I gratefully acknowledge permission to republish parts of previously published work, including:

al-Rai al-Am, for permission to reprint the cartoons and advertisement featured in chapter 2.

Sage Publications, for permission to reprint parts of an article, "The Internet and Public Culture in Kuwait," which appeared in *Gazette: International Journal for Communication Studies* 63, 2-3 (May 2001): pp. 187-201. Parts of this article are featured in chapter 3.

State University of New York for permission to republish parts of a chapter, "New Technologies, Old Culture: A Look at Women, Gender, and the Internet in Kuwait," which appeared in *Culture, Technology, Communication: Towards an Intercultural Global Village,* edited by Charles Ess and Fay Sudweeks (Albany: State University New York Press, 2001). Parts of this article form chapter 4 of this book.

The Journal of Computer Mediated Communication, which published "The Internet and the Politics of Youth Subculture in Kuwait" in March 2003. Portions of this article appear in chapter 5.

CHAPTER 1

———

The Internet in Global
and Local Imaginations

The parent or inventor of an art is not always
the best judge of the utility or inutility of his
own inventions to the user of them.
—Plato, *The Phaedrus*

The real revolution is in this latter and pro-
longed phase of "adjustment" of all personal
and social life to the new model of perception
set up by the new technology.
—Marshall McLuhan, *The Gutenberg Galaxy*

Plato, many centuries ago, warned us not to let inventors
judge the value of their creations. He told us that inventors "from a paternal love of [their] own children have
been led to attribute to them a quality which they cannot
have."[1] In the Internet's case, the inventors of the technology expected too little from their offspring. This situation of underestimation is understandable. After all, the
historical record suggests that the Internet was not born
under the noblest of circumstances. Harrison "Lee"
Rainie, director of the Pew Internet and American Life

1

project, captures the humble spirit of these earliest beginnings when he observes that:

> Internet communication began with a computer crash and an unmemorable twaddle. . . . Charley Klein, an engineer at the University of California, Los Angeles, froze his computer in 1969 when he began typing "L-O-G" (on his way to "L-O-G-I-N") to start the file transfer program. Programmers fixed the glitch quickly, and the file sharing began.[2]

The Internet has come a long way since then, becoming a basic tool of communication and information gathering for over half a billion people. Whereas for the first ten years of use, from 1969 to 1979, the Internet consisted of no more than 188 hosts, and a relative handful of users, by January 2003, 171,638,297 host computers and more than 600,000,000 users were linked to the network, demonstrating the Internet's phenomenal growth.[3] Even in the Internet inventors' most fantastical moments, such depth, speed, and breadth of sprawl were unlikely to be anticipated.

Over time, the tool has proven unimaginably useful to a myriad of different pursuits—economically, politically, and socially. Given the Internet's open architecture and acephalous organizational structure, the line between users and producers has blurred from the beginning, making the Internet a network of networks of utility, function, and meaning. As the Internet evolves, use patterns continue to be flavored by differences in class, gender, race, geographic location, national identity, age, sexual orientation, and other forms of individual and communal distinction. User characteristics, along with a host of other contextual factors, further complicate any general definition of what the Internet means, both for those using the technology and for the ways we live.

Such complexities in architecture, use, and implication have not stopped social scientists and others from attempting to empirically document the Internet's many meanings, and this book is no exception. Methods in the emerging field of Internet Studies range from "large n" surveys adopting a variety of statistical tools to more constructivist forms of virtual and real ethnography.[4] Some scholars use the Web as a research site as well as a method for extracting data (a field site; Web surveys).[5] A minority of scholars has used more traditional anthropological research methods to study online and off-line behavior, and in the case of this book, the relationship between them.[6] Another minority of scholars has studied Internet meanings in non-Western societies.[7]

This study carves a place for itself in the emerging field of Internet Studies by using "small n," ethnographic methods to study the relationship between context and Internet use in an Islamic country. Rather than try to study the Internet as a powerful, universally meaningful device, this study instead takes a more microanalytical approach by focusing on the Internet and its meanings for communities of users in Kuwait. What seems apparent more than anything else in Internet Studies is that the Internet is a much richer tool than anyone has yet been able to conceptualize or explain. Perhaps this is because, as Daniel Miller and Don Slater, two Internet ethnographers, observe: "The Internet is not a monolithic or placeless 'cyberspace'; rather, it is numerous new technologies, used by diverse people, in diverse real-world locations."[8] This book studies Internet practices in Kuwait as a way to show how local culture and politics shape the meaning of the global Internet.

This study is intended to link to more general attempts to conceptualize the Internet's global meanings. In the process, it provides an analysis of clusters of local Islamic Internet uses and meanings to counterbalance a

predominance of Internet theorizing taking place in North
America and Europe. As a team of Internet researchers
from the University of Toronto has observed:

> Social research about the Internet has followed
> the spread of the Internet itself. With the Internet
> born and raised in the USA, most research has
> been American. With Internet use increasing in
> other developed countries, research about their
> situations has been on the rise. However, there has
> been little research about how Internet use fits into
> the everyday life of developing countries. Further-
> more, international comparisons are almost non-
> existent.[9]

These circumstances provide significant incentive for
turning our gaze to countries such as Kuwait, countries
that possess a host of Internet active pockets of population and a range of cultural, political, economic, social,
and geostrategic differences, some of which help make
the Internet a distinct part of everyday life. The main
message of this book is that while the network links
global users and enables information to flow across borders and identities, location still matters in shaping online activities.

Before evaluating the ways in which the Internet matters to Kuwaitis, this chapter attempts to provide an
overview of the technology's evolution and examines some
of the values and tensions embedded in its design and
use. These original design principles enabled the Internet
to become adaptable to Kuwaiti preferences and imaginations. At the same time, this chapter situates a study of
Kuwaiti Internet practices within the growing body of Internet research, known collectively as "Internet Studies."
This chapter illustrates how the expectations attached to
the Internet in the twenty-first century go far beyond the

limited vision of the inventers of the technology, a process that fits a general pattern of technological evolution. We often see technologies designed for purposes they fail to achieve, or achieving purposes for which they were not designed. The unintended consequences of technological innovation are partly a product of the inability to fully grasp the social construction of a technology's meaning and purpose for users. The design of the Internet has enabled over time a distributed network of users to add value to the Internet in adapting the technology to their own needs, from new programs and pieces of Web architecture such as e-mail, newsgroups, and video conferencing to innovative Internet practices such e-commerce, chat, and blogs. The distributed, nonhierarchical nature of the network has made Internet adaptations and innovations in locations such as Kuwait central and meaningful to the Internet's growth as a global phenomenon.

Designing the Network, Distributing a History, and Empowering Users

The design principles of the Internet have made the technology open to adaptation and innovation from its inception. In fact, as examined later, were it not for this open architecture, the technology might not have emerged as a global phenomenon. Understanding this history and adaptability is crucial to grasping the ways in which the technology itself allows for the Internet to be given a particularly Kuwaiti meaning and implication.

Constructing a history of the Internet is a difficult task given the distributed nature of computing innovations that have, brick by brick, laid the foundations for what we now know as "the network of networks." Most people who study and read about the Internet have a very general picture of the Internet's evolution. Some begin with Paul Baran and the RAND Corporation's project in the late 1950s to envision and design distributed communications networks that would survive a nuclear attack. Since this project was never fully realized, a more common beginning is the Advanced Research Projects Agency Network (ARPANET), a project that began to take shape in 1962 under the direction of the US Department of Defense Advanced Research Projects Agency. The ARPANET was an actual network that linked a handful of universities and centers of scientific research through a type of distributed computing and information exchange, known as "packet switching." When the ARPANET was established, the goal of the project was not about enabling communications to survive nuclear war. Rather, this phase in the Internet's evolution materialized out of a desire "to mobilize research resources, particularly from the university world, toward building technological military superiority over the Soviet Union," and more specifically, "to stimulate research in interactive computing."[10] According to Manuel Castells, however, "to say that the ARPANET was not a military-oriented project does not mean that its Defense Department origins were inconsequential for the development of the Internet."[11] Rather, "The Cold War provided a context in which there was strong public and government support to invest in cutting-edge science and technology."[12] As explored more completely later, military values also influenced the design of the technology, especially the issue of survivability.

In 1973, adding a node to the ARPANET at University College of London created the first international link to the Internet. By 1982, due to the vision of Bob Khan and Vint Cerf's Transmission Control Protocol/Internet protocol (TCP/IP) suite, building upon the earlier vision of Paul Baran and Donald Davies, the Internet was given the technical capacity to evolve into the network of networks.

In 1983, Stuttgart and South Korea joined the ARPANET as a part of the network's international expansion. The key to building the global Internet was the adoption of an internationally shared and standardized protocol for packet transmission.

Phase two was Advanced Research Projects Agency (ARPA's) transfer of responsibility for managing the growing network to the National Science Foundation (NSF) in the early 1980s so the network could be opened up more freely to civilian use. This transfer was originally achieved by dividing the network into the civilian ARPANET and the government/military network, MILNET. By 1987, ARPANET was retired, and National Science Foundation Network the NSFNET took over as the backbone for the Internet, with former ARPANET sites linking to the NSFNET. Many Middle Eastern countries, and a host of others worldwide, gained their first taste of Internet access by linking local universities to the NSFNET. For example, in 1991, Tunisia was the first Middle Eastern country linked to the NSFNET. In 1992, Cyprus and Kuwait linked to the the NSFNET. In 1993, Egypt, Turkey, and the United Arab Emirates (UAE) established links to the NSFNET. In 1994, Algeria, Jordan, Lebanon, and Morocco joined the NSFNET. Universities and colleges linked to the network had to agree that access would not be used for commercial purposes.

Phase three was the decision taken by the NSF and the U.S. Congress to open the Net to commercial service providers. Opening it to commercial competition and use resulted in the Internet's spread as a mass technology. High-profile moves online also stimulated Internet growth, drawing business and media attention and creating the beginnings of a public culture of the Internet. For example, in 1992, the World Bank established its presence in cyberspace (*http://www.worldbank.org*), and in 1993, the White House went online (*http://www.*

whitehouse.gov). Also key in encouraging a growing public awareness of the Internet was the creation of software to make the Internet more user friendly for nontechnical communities. In 1995, commercial dial-up systems such as America Online (AOL) and CompuServe began providing commercial services designed for the general public. The voice of AOL's "You've got mail" would become an important icon in American popculture, from Sesame Street to a feature film by Nora Ephrim, entitled *You've Got Mail,* starring Tom Hanks and Meg Ryan. In 1995, Netscape, a browsing program created by Mark Andreeseen and a team of other young computer scientists, went public. The Netscape browser revolutionized the way we access information on the network, making Web surfing graphic, more user-friendly, and ultimately more commercialized.

From the beginning, the inventors of the Internet worried that normal phone lines would not be able to handle growing amounts of network traffic. As the Internet has developed into a mass-based technology, engineers in both the private and public sector have created plans to build a bigger, better, faster pipeline through which data will flow, and such efforts constitute a fourth phase in the Internet's evolution. Examples of such efforts include the FLAG project (Fiber Optic Link around the Globe), the Internet 2 project, and projects such as Teledesic, those that attempt to make the Internet wireless and more widely and cheaply accessible. The goal of these efforts is to establish a network that will allow instant video over IP and a more efficient transfer of larger bodies of digital information, including that which is needed for ease with video conferencing.

The bigger picture behind the Internet's evolution is much more complicated. A global cast of characters worked in a networked way to build bits and pieces of software to link hardware; to build protocols and standards that would enable data to flow globally, safely, and uniformly; and to establish a cluster of distributed networks and backbones around the world, forming the channels through which bits and bytes flow. Eventually these elements meshed together human minds and collective technological evolution to build and shape the network of networks that we know today as the "Internet."

Early Visions

The vision behind the project can be traced all the way back to the mid 1940s when Vannevar Bush conceived of an Internet-like device that he called a "memex." A memex, according to Bush,

> is a future device for individual use . . . a sort of mechanized private file and library. . . . A memex is a device in which an individual stores all his books, records, and communications, and which is mechanized so that it may be consulted with exceeding speed and flexibility. It is an enlarged intimate supplement to his memory.[13]

Another early visionary was Joseph C. R. Licklider, the first director of the ARPA's Information Processing Techniques Office. In 1960, he wrote what would become a highly influential paper, "Man-Computer Symbiosis," in which he observed:

> The hope is that in not too many years, human brains and computing machines will be coupled together very tightly, and that the resulting partnership will think as no human brain has ever thought. . . . Those years should be intellectually the most creative and exciting in the history of mankind.[14]

These early visions of a technology that was part library, part entertainment zone, and part extension and enhancer of human brain work have to some degree been fulfilled by the emerging history of the Internet. Although these mythical beginnings inspired early inventors to dream big, more fundamental to the venture was a whole host of more pragmatic and practical scientists and academics who made the dream a reality through their problem solving, ingenuity, and cooperation.

Although much of the initial impetus for the project was American, the cast of those involved in building the Internet as we know it today includes a host of Europeans, including two important Brits, Donald Davies, who invented his own idea of packet switching, contemporaneously, but with no knowledge of Paul Baran's work at RAND, and Tim Berners Lee, a Brit working in Switzerland at CERN, a nuclear physics laboratory in Geneva, who invented the World Wide Web. If Web content and local networks are included in the definition of the Internet, then a crew from more than 241 countries has participated.

History in a Nutshell

The complex and twisted networks of engineering and computing breakthroughs that led to our modern-day Internet have not kept a handful of observers from elegantly capturing history in a nutshell. For example, Neil Randall, in his book *The Soul of the Internet*, states:

> The soul of the Internet is, first of all, the people who envisioned it, developed it, and hammered and cajoled it into its present global shape. . . . The visionaries have been few, while the developers and

builders have been many. . . . For every Vannevar Bush and his memex system, T. H. Nelson and his Zanadu fixation, and J. C. R. Licklider and his libraries of the future, there have been thousands of engineers, computer scientists, psychologists, and researchers from other related fields laying the groundwork on which these visions might build. . . . Whatever the reasons for the Internet's beginnings, whether it was a scientific initiative with a military payoff or a military initiative with a technological payoff, the community of builders saw a job that needed doing and kept working at that job until it reached another stage of completion. Together they created what promises to be the most significant technology of our time.[15]

Adding a place for the netizens and everyday users of the Internet in the co-evolution of the network, Randall continues, "The soul of the Internet is still the collective soul of its builders, but we—all of us who use it—are now the ones who are building it. Whether we add new content or simply create a demand for this content, we are responsible for what it [the Internet] will become."[16]

Janet Abbate has written one of the best histories of the Internet. One of the qualities distinguishing her account is that hers is a social history, focusing on the individuals and visions that built the network as much as on the cultural contexts and human necessities that helped shape the network's design. She captures the collective spirit behind the Web when she observes the following:

The cast of characters involved in creating the Internet goes far beyond a few well-known individuals, such as Vinton Cerf and Robert Khan, who have been justly celebrated for designing the Internet architecture. A number of ARPA managers contributed to the Internet's development, and military agencies other than ARPA were active in running the network at times. The manager of the ARPANET project, Lawrence Roberts, assembled a large team of computer scientists that included both accomplished veterans and eager graduate students, and he drew on the ideas of network experimenters in the United States and in the United Kingdom [especially the work of Donald Davies of the UPL on packet switching]. Cerf and Kahn also enlisted the help of computer experts from England, France, and the United States when they decided to expand the ARPANET into a system of interconnected networks that would become known as the Internet.[17]

In addition to the engineers and computer scientists who envisioned, built, and expanded the network, Abbate also highlights the role of other key institutions in evolving the network when she suggests that "telecommunications carriers, vendors of network products, international standards bodies," and the National Science Foundation all helped build a globalized Internet.[18] Tim Berners Lee's invention of the World Wide Web made a major contribution to the global expansion of the Internet by making it more user-friendly and browsable. Had the U.S. government not contributed large amounts of financing to the ARPANET and NSFNET projects, then the Internet might not have ever emerged. Lack of sufficient capital kept British scientists from independently inventing the Internet beyond theoretical prototypes. A complex web of individuals, institutions, financial and technical resources, and politics illustrates that the "history of the Internet is not, therefore, a story of a few heroic inventors; it is a tale of collaboration and conflict among a remarkable variety of players."[19]

Values, Culture, and User Imaginations in Inventing the Internet

The design principles of the Internet, particularly its distributed links and nonhierarchical organization, enabled the co-evolution of the technology and has assured a large place in the Internet's history for users. The distributed network of producer-users also has meant over time that context of use remains important in shaping the Internet's meaning. Abbate claims that the most important values expressed in the design of the Internet include the network's "informal, decentralized, user-driven development."[20] The military and its values also played a role in designing the Internet. Abbate tells us that "ARPANET and the Internet favored military values, such as survivability, flexibility, and high performance over commercial goals such as low cost, simplicity, or consumer appeal."[21]

Although commercial goals were not part of the network's original design, the flexibility of the Web is clearly demonstrated by the hegemony of business uses of the Web in the twenty-first century. Even though it took more than thirty years for business values to shape the Web, the Internet would not be what it is today without them. Some of the commercial values shaping the Net include the idea that having a presence in cyberspace is required for doing business. Also fundamental to the creation of massive amounts of Web content, as well as an incentive for drawing individuals online, was the notion that it made good business sense to give away information on the Web, to create Web site "stickiness" or attractability. The ideas of e-commerce, telecommuting, and the creation of individualized business services such as e-banking, e-trading, Web-based personalized travel features such as those provided by Expedia and Orbitz all helped contribute to the spirit and culture of the Internet, enhancing individual user capabilities to master complex previously professionalized services. Corporate culture and business practice have been reshaped by the enhanced collaborative enterprise characteristic of the knowledge economy where companies often pool intellectual capital for technological and business innovation, in part because the Web makes it possible.

In addition to military values embedded in the network's original design, academic values also were present in shaping the network, because many of the early inventors were professors or graduate students working with ARPA as civilian contractors. Abbate explains: "The group that designed and built ARPA's networks was dominated by academic scientists, who incorporated their own values of collegiality, decentralization of authority, and open exchange of information into the system."[22] This collegial spirit of sharing information and evolving nonhierarchical forms of information exchange and engagement have aided in devolving power to innovate to individual netizens.

Users Design the Network of Networks

E-mail is a perfect example of inventors' vision being superceded by user preferences. Originally the Internet was not envisioned as a communications device but rather as a computing tool, designed to enable distributed networks of computers to run shared software. Yet even though the Internet was not designed primarily for e-mail, the exchanging of messages among users was, and still is, one of the most common uses of the Internet. Abbate reinforces this view when she observes the following:

> The very notion of what the Internet is—its structure, its uses, and value—has changed radically over the course of its existence. The network

was not originally to be a medium for interpersonal communication; it was intended to allow scientists to overcome the difficulties of running programs on remote computers. The current commercially run, communication-oriented Internet emerged only after a long process of technical, organizational, and political restructuring.[23.]

Ray Tomlinson created the first e-mail program, in the early 1970s, while employed as an engineer at Bolt, Beranek, and Newman. It would take another twenty years of evolution to make e-mail a regular feature of everyday life. Even when Internet use was confined to a relatively exclusive community of users, email was the dominant feature of online behavior. For example, 75 percent of all ARPANET traffic from 1969 to 1973 was e-mail.[24] Barry Wellman and Caroline Haythornthwaite observe that as of 2000, over 80 percent of Internet use is for e-mail.[25] E-mail is consistently the number-one reason given in user surveys for being online, according to the research of J. E. Katz and P. Aspden.[26] A study by UCLA's Center for Communication Policy estimates that in the United States alone, in 1998 more than 4 trillion e-mail messages were sent; moreover, 42 percent of Americans check their e-mail at least once a day (2000 est.).[27] The predominant character of the Internet as a communications tool led Michael Strangelove to observe, "The Interent is a community of chronic communicators."[28]

The inventors of the Internet did not design the technology to be a worldwide phenomenon or a mass technology supporting day-to-day communication strategies. The values they embedded in the technology, however, including the military values of survivability and flexibility and the academic values of collegiality and the open exchange of information, carved out the necessary spaces in which users could take an active part in designing the technology to meet their needs and wishes. The adaptability of the technology has made this revolution in communications and ways of living fruitful ground for social scientific explanation. We now turn to the emerging field of Internet Studies.

Internet Studies: Key Assumptions and Questions

As suggested by the previous section, the Internet has been evolving for approximately forty years. The last ten years of its development, however, have been the most important to social scientists. It took thirty years of evolution for the technology to become what John Perry Barlow calls "the most transforming technological event since the capture of fire."[29] This statement may be what Steven Woolgar calls a bit of "cyberbole."[30] But there can be no mistake in acknowledging that while the Internet was once "an obscure academic playpen," it has now become a key tool in the practice of everyday life.[31] The fact that this statement is true for some societies more so than for others has not curbed the perceived importance of having access to the network. As the Internet increasingly becomes an extension of human potential, we are encouraged to consider its role in shaping our future as a world community. This book is a part of that process.

A kind of "Barlow-like" enthusiasm for the Internet reverberates through much of the social scientific literature on the subject. Just highlighting a handful of the debates reveals a rich canvas of futurist imaginary. There is an assumption built into much of Internet Studies that "The Internet changes everything," to borrow a phrase from Bill Gates, one of the builders of this revolution. In less sensationalistic and more specific terms, Jessica Mathews, president of the Carnegie Foundation, grounds Gates's hyperbole by observing, "As the information revo-

lution rapidly unfolds, Internet use is profoundly affecting governments, corporations, and societies around the
world." At the same time, she notes, "many of these effects, while widely assumed to be significant, have yet to
be fully explored."[32] This sense of a rapidly unfolding
technological revolution matched by equally profound
changes in human behavior is what draws social scientific innovators to the emerging, interdisciplinary field of
Internet Studies. As suggested by Matthews's observation, the Internet is creating a kind of scholarly gold rush,
where staking a claim in this emerging intellectual universe provides the rich rewards and satisfactions of being
on a new frontier, where Internet-inspired power struggles, economic opportunities, rules and regulations, and
customs and everyday practices are being transformed
before one's eyes.

Another assumption is that the Internet is able to
change everything because it is becoming a regular feature of everyday life. Katie Hafner, in describing the natu-
ralization of the Internet, observes: "No one owns it. And
no one in particular actually runs it. Yet more than half a
billion people around the world rely on it as they do a
light switch."[33] Castells embellishes this view of the fundamental contribution of the Internet to digital civilization
when he observes, "The Internet is the fabric of our lives.
If information technology is the present-day equivalent of
electricity in the industrial era, in our age the Internet
could be likened to both the electrical grid and the electric
engine because of its ability to distribute the power of information throughout the entire realm of human activity."[34] In illustrating a key link between technological
change and social-institutional reorganization, Castells
goes on to argue that "as new technologies of energy generation and distribution made possible the factory and
the large corporation as the organizational foundations of
the industrial society, the Internet is the technological

basis for the organizational form of the Information Age: the network."[35] In other words, change the ways in which information flows and how we communicate, and change the way we work, live, and play.

A third assumption behind Internet Studies is that the scope, speed, and depth of changes are unprecedented. Frank Webster, Britain's best-known analyst of the information society, notes that in the Internet age, "the rhythms of everyday life, our daily experiences, our routine schedules, are being transformed in ways which, by any historical comparison, are remarkable."[36] While the Industrial Revolution took more than 100 years to really come to fruition, to change the patterns and values of life for the mass of society, Webster observes of the information revolution, "We need look back scarcely a decade to trace the spread of information technologies through office, home, and education. In the space of a very few years, digitalization has become a routine part of life."[37] The U.S. Information Infrastructure Task Force reported the following comparisons between the spread of the Internet and other communications technologies. In terms of reaching a concentration of 50 million users, it took radio thirty-eight years, TV thirteen years, and the Internet only four years (once made commercially available) (*http.//www.ecommerce.gov/emerging.htm*). The Mosaic Group's "Global Diffusion of the Internet Project" in 1998 observed of the Internet that, "it has been the most rapid and extensive diffusion of advanced technology in history."[38]

Just because the Internet is becoming an important part of everyday life for the majority of people living in North America does not mean that the same is true for the rest of the world. In the Middle East and North Africa, users sometimes fail to constitute 5 percent or even 1 percent of society, as discussed in more detail in chapter 2. A significant body of Internet Studies is dedicated to the analysis of the causes and implications of the digital divide separating those societies that have rapidly absorbed Internet technologies and those that have not. A guiding assumption of this literature is that the Internet is a site of empowerment, economically and in terms of social capital generation. Therefore, the world cannot afford to just stand by and watch as the information poor become increasingly marginalized from the spoils of digital civilization. This literature often works in tandem with theories of international development and tends to have a heavy dose of policy analysis built into it.[39]

In describing what is so special about the Internet, Alexander Bard and Jan Soderquist, authors of the controversial book *Netocracy*, highlight another common assumption in Internet Studies, and that is the technology's uniqueness. They observe:

> The Internet is something completely new: a medium in which virtually anyone, after a relatively small investment in technical equipment, and with a few simple actions, can become both a producer and a consumer of text, images, and sound. It is hard to think of anything more empowering: on the net we are all authors, publishers, and producers; our freedom of expression is as good as total and our potential audience limitless.[40]

Others have stressed the uniqueness of the Internet by contrasting it with "old media."[41] The Internet is new media in that it is a "many to many" form of broadcasting information and entertainment. It also is new because of convergence issues. That is, the Internet is at the same time a communication tool, such as a post office, a telephone, and a fax machine. It also is a gigantic library, a chain of places to shop, and a television, a newspaper, a radio, a place for gaming and role-playing. The convergence of all of these techniques makes the Internet unique. In part the technology itself is responsible for the importance of novelty in the digital age. Randall, in *The Soul of the Internet notes:* "It goes almost without saying that, on the Internet, new is everything. Simply, there has been no other medium in our history that has demanded as much as the Internet that things be new, that the old does not count."[42] Castells adds to the debate when he observes that the Internet "is something special." This is because:

> New uses of the technology, as well as the actual modifications introduced in the technology, are communicated back to the whole world, in real time. Thus, the time span between the processes of learning by using and producing by using is extraordinarily shortened. . . . This is why the Internet grew, and keeps growing, at unprecedented speed.[43]

There also is a body of scholarship that attempts to counter the notion of the Internet's uniqueness by linking the technology to a long history of innovations in communications.[44]

Just as social context helps shape the Internet's meaning, so too is Internet scholarship influenced by location. We see such local values expressed in relation to the Internet's meaning in Bard's and Soderquist's passage quoted earlier. They express European contextual values in their use of the phases "virtually anyone," "after a relatively small investment," "after a few simple actions," and "our freedom of expression is as good as total." Perhaps these statements are true for Europe, where this text was produced and the Internet is more common, but for those on the losing side of the digital divide, the Internet is complexly out of reach. Moreover, for those who access the Internet in authoritarian countries, the authors' emphasis on total freedom of expression is naïve, if not dangerously exaggerated. Since the context of research seems to shape views and values of Internet theorizing, scholars of Internet studies are increasingly committed to the exchange of local Internet knowledge among the emerging global community of Internet researchers.[45]

Another debate in Internet studies concerns the division between techno-determinism and social constructivism. Some Internet researchers assume that the Internet is in and of itself a force for change, that it determines the course of history. There is an assumption that "technology plays its own hand" built into this approach to Internet Studies.[46] The techno-determinist approach to Internet Studies draws inspiration from the rich literature in Science Technology and Society (STS), some of which operates from the perspective that "history shows that every major new technology has, for better or for worse, done its own thing; completely independently of what its originators had imagined."[47] In this sense, technology determines social outcomes. For some, it is not that the Internet determines anything directly but rather "by creating patterns of reliance and dependency through which our lives will be indirectly and irrevocably reshaped" determines the course of humanity.[48] Thus it is the dependency on the tool that determines the shape of human activity.

In contrast to the strong determinist literature, there exists the softer, more intuitive social constructivist literature that views technology "as a way to conduct meaningful social practice" to the degree that "technology does not exist in a 'pure objective form' outside of the context of social practice."[49] From this perspective, the meaning and implication of the tool are found in society's use of technology, and this is why we see the same tool, the Internet, having different meanings in a variety of contexts.

A synthesis position also exists in Internet Studies that takes a middle ground between techno-determinism and social constructivism and is best exhibited in the foundational work of Castells. Take, for example, his observation that "rather than analyzing the impact of the Internet on society, the key issue is to understand the effect of society on the Internet."[50] This statement makes Castells look like a social constructivist. Evidencing a synthesis of techno-determinism and social constructivism, however, Castells continues, "the Internet is not just a tool, it is an essential medium for the network society to unfold its logic. This is a clear case of co-evolution between technology and society."[51]

Most scholars in the field of Internet Studies take the middle ground between techno-determinism and social constructivism. They recognize how the Internet enables certain practices and relationships that were otherwise not possible, while at the same time reserving a space (sometimes larger than others) for social and contextual variables that help shape how new possibilities and ways of interacting are imagined and enacted via the Internet, or not. Sorting through the ways in which the Internet determines and is determined by a host of factors remains an important debate in Internet Studies.

First-generation Internet research tended to be more techno-deterministic, whereas as Internet scholarship evolves, context of use becomes a more stringent variable. This trend "reflects the recognition that users are not simply the passive recipients of technology but that they are active and important actors in shaping and negotiating its meanings."[52] This softening of techno-determinism has been clearly illustrated, for example, in the literature on the Internet and democratization. Kalathil and Boas, in their six-country study of Internet use in authoritarian settings, are encouraged by empirical evidence to tame "blind optimism" that the Internet promotes democratization. In explaining why the Internet does not necessarily promote democratization in authoritarian states, Kalathil and Boas highlight a number of contextual constraints not previously entertained in the more general literature on the subject. They write:

> In most authoritarian regimes, the state has historically played a strong role in the development and control of ICTs and the mass media. This legacy usually translates into a dominant role in Internet development. While the academic or scientific community often takes the lead in early experimentation with Internet technology, state authorities are almost always responsible for guiding the broader diffusion of the Internet within national borders. In the course of doing so, they typically devise both technological and policy architectures that facilitate state control over the Internet.[53]

The case of the Internet and authoritarianism illustrates a common pattern in Internet studies where an expectation unsupported by empirical evidence yields to further scrutiny at the points of application, in actual authoritarian contexts, thus yielding to a more nuanced view of, in this case, the Internet and politics. A more detailed analysis of the idea of linking the Internet and democratization is analyzed in light of the Middle Eastern data in chapter 2.

Those who study and explain patterns and meanings associated with the Internet revolution are not daunted by the information overload, the lack of norms, the absence of an accepted canon of theories, the need to build data sets often from scratch, and the rapidity with which data sets are outmoded by the speed of the network's shape-shifting, but rather are nurtured by it. In Internet times, there is much space for intellectual play, as old conventions yield to new communication patterns and ways of living. It is this sense of being intellectually free, and intensely drawn to the puzzle and prospects of the Internet that motivates the field of Internet Studies.

Key questions in Internet research include the following partial list: Will the Internet play any role in making us richer, happier, or more secure, politically, economically, and socially? Will there or will there not be less war and more cooperation locally and globally in the Internet age? Will we or will we not wipe out poverty, neglect, homelessness, and globally unsatisfied need? Will we or will we not use digital technologies to stop global warming, better understand the human genome, solve diseases that have stumped us, and more adequately share basic health care with the world community, with our new knowledge pools and new delivery techniques? Will good governance and democratization be more common, less common, or remain unchanged in the Internet age? Will more people have a say in what good governance means and how it is implemented as a result of Internet access? Will we be more apt to voice our opinions and to learn from those with whom we disagree as a result of new forms of information and conversation made possible by online environments? Or instead, will we simply build greater solidarity with like-minded individuals and communities (hate groups, for example) and put at risk deliberative democracy? What happens to race, class, gender, ethnicity, sexual orientation, and all of the other identity layers used to promote and exclude individuals? Does the digital age mean more opportunities for the previously marginalized, or less? Will the Internet result in economic growth opportunities for developing countries? What about questions of access and control? Is state sover-

eignty at risk as information, capital, and people flow
with increasing speed and scope across or beyond national boundaries? Who gets to govern cyberspace? What
are the major issues demanding our attention in terms of
regulation? Which regulations work best to encourage the
proper use of the Internet and to make the network more
secure for business, individual use, and government service distribution? What will be the results of a more carefully regulated Internet? Should we be worried about surveillance, privacy, and individual rights? Should we trust
the network, those shaping the network, and/or should
we be vigilant in assessing its almost invisible creep into
our daily lives? How will or should the Internet be used in
education? In what ways does the Internet enable new
kinds of crime and security risks? What is distinct about
digital civilization? What do we not know about the digital
age, and what should we watch, in terms of humanistic
concerns? These are just a prominent few of the many
questions underlying Internet Studies.

What We Know about the Internet

While it was once fashionable to speculate that the
"The Internet changes everything," with nearly ten years
of empirical research in the emerging field of Internet
Studies accumulating, many of our original expectations
have now been more carefully investigated and adjusted
to fit the data. Because the claims about the Internet
were in the initial stages so fantastic and speculative, a
number of projects emerged to put the hype to the test.
Prominent among the large survey attempts at docu-
menting the Internet's meaning is the Pew Internet and
American Life project. The director of the project, Harrison "Lee" Rainie, explains that the project was designed to test "an orgy of extravagant claims about the

revolutionary power of the Internet."[54] The findings were mixed. Rainie explains:

> In the course of doing telephone interviews with more than 60,000 people during the past 30 months, the Pew project learned that Internet use is helping Americans to share and acquire knowledge, make important health care decisions, deepen and extend their social networks, access cultural material, probe new corners of the planet, pursue their passions and hobbies, become more productive, gather up more consumer information, and entertain themselves more vividly.[55]

Arguably one of the most important findings of the Pew survey for this book, and for those committed to a social constructivist approach to the Internet, is that even within the borders of the United States, "context matters a great deal in the way in which people use the Internet and how they feel about it"[56] (*http://www.pewinternet. org*).

Speaking more generally, the major findings of Internet Studies can be summarized as follows: We know not to believe the hype, to be skeptical of predictions regarding the Internet, when not supported by empirical research. We know that given the importance of context and user characteristics, and the persistent viability of place as a unit of social organization even in the virtual age, it is difficult to generalize about the Internet's meaning beyond specific locations. Like all knowledge, Internet knowledge is bounded. The way in which the Internet is used, or not used, "is defined in large part by their [users'] local everyday lives, the social, political, economic, and cultural environment in which they live, and by the ways in which they appropriate these technologies."[57]

In terms of democracy and the Internet, we know to

be cautious in our attempts to link the two. In authoritarian contexts, the Internet is slow to wrought change. Just because some citizens have access to uncensored information via the Net does not mean that human rights, institutions for civic participation, and freedom of speech and association automatically appear.[58] In advanced democracies, the Internet is expected, on the contrary, to undermine collective associational life and to interrupt the deliberative process, thus putting democracy at risk—but, at the same time, new forms of activism are enabled by the network—such as the global World Trade Organization (WTO) opposition movement.[59]

In terms of sociability, being online has not, as expected, led to social breakdown and alienation as virtual contact increases. Instead, studies suggest that "the proliferation of the Internet means that people communicate more, not less. Internet use does not replace other forms of contact."[60] In addition to not replacing face-to-face contact, the Internet also "expands the reach of social networks that occur through the enhancement of the individual user's exercise of communicative power."[61] Thus the Internet, like building forms of globalized democratic politics not tied to any particular location, also expands human relationships beyond normal forms of everyday social interaction, building personal networks across vast spaces.

In terms of economics, we know that information technology and the Internet-do not automatically build wealth and economic productivity. In fact, a recent study of Internet-led economic development in Asia reveals that only those societies that produce information technology (IT) and also consume large amounts of it, reap significant economic growth.[62] Moreover, the poor seem to be getting poorer in spite of, or perhaps because of, the global spread of the Internet.

We also know more and more each day about general

use patterns, including data from professional research and marketing firms. One of the best portals for marketing research is NUA Surveys (*http://www.nua.com*). Based in Ireland, this company's Web site changes daily, giving access to data arranged topically, demographically, and regionally in terms of Internet impact. On February 18, 2004, for example, in randomly surfing the database, resources yielded the following:

- Twenty-four percent of children surveyed in Ireland, Belgium, the United Kingdom, and Greece have accidentally encountered pornography while online; 48 percent of Irish children said that they are never supervised while online; and 86 percent of Irish youth who use chat rooms were asked to meet in person someone they met on-line. All of these factors reveal security risks for children who use the Internet.[63]
- A report on Internet use in India reveals that cyber cafes are a common point of entry for young Internet users. Nearly 50 percent of school children surveyed access the Internet at these cafes. The same study also notes that the most popular web sites with Indian youths are sports and entertainment sites.[64]
- A survey on Internet use in the Arab world predicts that by the end of 2005, there will be 25 million Internet users in the region. Egypt is expected to have the most Internet users in the region by 2005 (as a percentage of regional users, not as percentage of local population).[65]

While our collective understanding of the Internet and its global meaning is aided by survey research, what is less commonly available are ethnographic data on Internet use in general, and particularly in the developing world. Hein ends her study *Virtual Ethnography* with this same conclusion when she observes that what is needed is more research on "the ways in which the Internet is interpolated into the concerns of particular local contexts of use."[66] In particular, she advocates for research that focuses "on the off-line contexts within which the Internet is used, tracking the ways in which the Internet is used in different contexts and the transformations which Internet content goes through as it passes from online to off-line contexts."[67] The chapters that follow answer this call for more contextually driven ethnographic Internet research in an attempt to expand global understanding of the Internet and its local meaning in the Islamic world.

Kuwait and the Internet: Building upon and Going beyond Three Themes in Internet Studies

Three debates in Internet studies were the starting points for this book, even though the analysis was compelled by empirical evidence gathered in Kuwait, or a lack thereof, to abandon these organizing principles. The particulars of Kuwaiti Internet culture provide an enlightening null hypothesis in three main areas of Internet research. First, the data contribute to the growing body of literature that is skeptical about the democracy dividend attached to having access to the Internet. Second, the data illustrate that Internet access does not necessarily promote economic growth and transitions to a more market-based economy. Third, the data show that although there is a tendency in Internet Studies to associate access with improving the power and freedom of individuals, such is not necessarily the case in politically and religiously conservative, if not authoritarian, contexts. Instead, this study of Internet practices in Kuwait chronicles the ways in which the Internet matters to three branches of Kuwaiti—society, women, youths, and Islamists. These three branches of society showed the most enthusiasm for the technology and seemed to be using the tool in the most culturally illustrative and innovative ways. Internet use within these three communities illustrates how different contextual variables shape the meaning and use of the Internet.

Chapter Outline

The chapters that follow help explain the development, character, and importance of Internet culture in Kuwait: Chapter 2 provides an overview of the development and importance of the Internet in the Middle East and analyzes the specifics of Kuwaiti Internet culture against the general regional and international backdrop; Chapter 3 looks at many contextual factors that help shape the configuration of Kuwaiti Internet practices; Chapter 4 examines women's access to and use of the Internet in Kuwait; Chapter 5 looks at youth subculture and the Internet in Kuwait; and Chapter 6 considers ways in which the Internet is used to promote Islamic awareness in Kuwait. The Conclusion interprets the findings regarding Kuwaiti Internet culture in terms of the technological and epistemological challenges of Internet Studies. Throughout this analysis, we see that the Internet does not so much determine a country's destiny, but rather it lends itself to the expression of a society's needs and values. In the process, of course, those needs and values are subject to their own renaissance, for in the Internet age, as suggested in the chapters that follow, the possibilities for new forms of social interaction are enabled by processes of fundamental technological change.

CHAPTER 2

———

The Internet in the Middle East, Kuwait, and Beyond

The Internet is growing and spreading more slowly in the Middle East and North Africa than any other place in the world. This chapter examines some of the reasons this might be so. It then looks more specifically at aspects of Internet developments in Kuwait. Kuwaiti Internet practices are examined against the backdrop of social scientific explanations regarding the expected implications of the Internet for the region. Internet practices in Kuwait do not necessarily conform to expectations for enhanced democratization, expansion of the private sector via the knowledge economy, and accentuated capacities for individual self-expression, both online and off-line. This does not mean, however, that Kuwaitis have failed to construct their own local versions of Internet culture and practice. On the contrary, the flexibility and adaptability of the Internet have enabled the emergence of a distinctly Kuwaiti Internet culture, as explored more completely in this chapter and in the chapters that follow.

Regional Connectivity

There are an estimated 5 million Internet users in the Middle East (not including North Africa), and nearly half of all users in the region reside in Israel. On the continent of Africa, there are an estimated 7 million Internet users, and 5 million of these reside in South Africa. If we total all Internet users in the Muslim majority societies of the Middle East and North Africa, we have 4,902,200 Internet users. It was estimated in August 2002 that only 2.54 percent of the population of the Middle East and North Africa (MENA) region has access to the Internet.[1] Moreover, Internet users here constitute only .75 percent of the 600,000,000 Internet users worldwide. In general, citizens of the Gulf Co-operation Council (GCC) countries have more access to the Internet than the rest of the Muslim Middle East.

The highest concentrations of Internet users per capita in the Arab world are found in the United Arab Emirates (36.79% per capita penetration, December 2001 est.), Bahrain (21.36% per capita penetration, December 2001 est.), Lebanon (11.22% per capita penetration, August 2002 est.) and Kuwait (8.91% per capita penetration, December 2001 est.). The lowest concentrations of Internet users per capita are found in Iraq (.05% per capita penetration, December 2000 est.), Yemen (.09% per capita penetration, December 2001 est.), Sudan (.21% per capita penetration (2001 est.), Libya (.24% per capita penetration, March 2001 est.), Algeria (.57% per capita penetration, March 2001 est.), and Egypt (.85% per capita penetration, December 2001 est.). In the world community, the highest concentrations of Internet users per capita are in the United States (72% of the population), Canada (62% of the population), and South Korea (53% of the population).[2] The largest amounts of data (bits and bytes) transferred in and out of the MENA region

Internet Users in the MENA Region (2001)		
Country	Number of Users	% of Population
Algeria	180,000	.57%
Bahrain	140,200	21.36%
Egypt	600,000	.85%
Iran	420,000	.63%
Iraq	12,500	.05%
Jordan	212,000	3.99%
Kuwait	205,000	8.91%
Lebanon	300,000	11.22%
Libya	20,000	.24%
Morocco	400,000	1.28%
Oman	120,000	4.42%
Palestine	103,000	3.03%
Qatar	60,000	7.59%
Saudi Arabia	570,000	2.5%
Sudan	70,000	.21%
Syria	60,000	.35%
Tunisia	400,000	4.08%
UAE	900,000	36.79%
Yemen	17,000	.09%
Total	4,902,200	2.45%

Figure 2-1. Internet Users in MENA Region (2001)

occurs in Egypt.[3] Some have argued that the amount of data transferred in and out of a country is a more robust indicator of Internet use and impact than the number of users per capita. Abdalatif-al-Hamad, chairman of the Arab Fund for Economic and Social Development, estimates that "In Europe, investment in IT is 1,000 times more than it is in the Arab world."[4] Differences in investment in the IT sector could indicate one reason growth in Internet use is latent in the MENA region.

As suggested by comparing Figure 2-1 and Figure 2-2, low numbers for per capita Internet use (generally determined by the number of Internet subscribers

Country	Number of Internet Cafes
Bahrain	90
Egypt	400
Iraq	50
Jordan	500
Kuwait	300
Lebanon	200
Libya	700
Oman	80
Palestine	60
Qatar	80
Saudi Arabia	200
Sudan	150
Syria	600
Tunisia	300
UAE	191
Yemen	120
Morocco	2,150
Algeria	3,000

Adapted from Madar Research Group.

Figure 2–2. Internet Cafes in the Middle East

multiplied by the estimated number of users per machine, usually a factor of 2–4) correspond positively to high numbers of Internet cafes in the region. For example, in Algeria, there is an estimated .57% Internet penetration, judged by the number of Internet accounts, but a high availability of Internet cafes, an estimated 9.52 cafes per 100,000 inhabitants. Similarly, in Libya, where per capita Internet penetration is estimated to be .24 percent, there are an estimated 13.21 Internet cafes per 100,000 inhabitants. Given the difficulties in determining what percentage of the population has access to the Internet via Internet cafes, it is hard to judge the mass appeal of the Internet. If in places such as Algeria and Morocco there is enough public demand for the Internet to support a 2,000 or 3,000 Internet cafes, then the Internet must be more important than indicated by conventional estimates. The large number of Internet cafes in the MENA region suggests that the Internet may be more of a mass technology than currently documented. Research on the role of cyber cafes in expanding public access to the Internet in other contexts suggests that such locations are places where people can overcome "issues of awareness, anxiety, and the need for new skills."[5] Further research on Internet access in cafes is important if a more complete regional picture of public use of the Internet in the Arab world is to be drawn. Access to the Internet in cafes could parallel access to newspapers in coffee houses during the late 19th century. It was difficult in the late 1800's to determine the public influence of newspapers in the Near East. Low numbers for circulation masked the fact that newspapers were shared among literate and non-literate audiences at coffeehouses, making them more of a mass media than circulation figures indicated. The same may be true for the Internet.

Another possible factor accounting for low Internet connectivity in the region is the rapid spread of mobile phones. In some cases, as in Egypt, the diffusion of mobile technology is more than double the spread of the Internet. In most countries in the Arab world, for approximately $100 one can purchase a mobile phone, a sim card, and one month's limited phone service. For additional months of service, one can purchase a pre-paid phone card. Moderate local use costs about ten dollars a month. There are also significant secondhand phone markets in most locations, where phones can be purchased for approximately 50 percent of their original cost. In the Middle East, unlike in the United States, the mobile phone customer only pays for outgoing calls. Receiving even international calls is free for the account holder. In some of my interviews in Kuwait and elsewhere, I was told repeatedly that mobile phone technology is more culturally compatible with Arab communication patterns than computer technology. The explanation was that a mobile phone is primarily a communications device, and since Arabs communicate daily with friends, family members, and business associates, they are highly attracted to 24/7 communications access provided by mobile phones. Computers, on the other hand, are viewed primarily as a computing or word processing device and only secondarily as a communications device. Most people in the region cannot afford mobile computing, such as a laptop, which costs up to twenty or thirty times as much as a mobile phone, thus the computer is not a 24/7 communication device. Moreover, there are more cases of friends and family having a mobile phone of their own than there are cases where both parties have computer access. In Egypt, for example, an Internet cafe manager in Dahab, explained that he had no reason to use the Internet because none of his friends or family members had access. With whom would he communicate? The *Arab World Competitiveness Report* claims that low Internet connectivity figures are a result of the fact that people regionally are not as comfortable using new technologies such as computers.[6] The fact that people regionally have easily learned to use mobile phones on a mass scale calls into question this assessment. It could be that computers are just not as adaptable to local needs and interests. Moreover, it is possible to send SMS text messages and e-mail from a mobile phone, thus negating, the need for a separate Internet service provider and a computer. Much greater attention to the everyday communication and computing patterns of the Arab world is necessary if we are to better understand connectivity, culture, and communication in the region. This is especially true in light of the fact that mobile phones in the Middle East are used to mimic some of the most common Internet practices—e-mail, chatting, and obtaining basic information such as news headlines, stock quotes, and weather reports. Also interesting, and worthy of further exploration, is the fact that mobile phone subscriptions in the region swelled to 30 million in June 2003, which is more than six times the number of Internet subscribers.[7]

The fact that connectivity in the Gulf is in some cases thirty times (per capita) more than it is in countries in North Africa and the Fertile Crescent suggests that economics also may influence Internet connectivity. Oil wealth makes Internet access more affordable for even economically average citizens in the Gulf, whereas in the rest of the Arab world, for the majority of citizens, Internet access is too expensive to be a part of everyday life. In Jordan, for example, where roughly 4 percent of the population has Internet access, the average wage of a government employee is 50 JD per month. In an Internet cafe, the most common way to access the Internet in Jordan, one hour of Internet use costs, on average, 1.5 JD, or about $2.10. The same one-hour fee for Internet use could easily buy a family of five lunches at a falafel restaurant. Internet access at home (which presupposes having access to a computer and a telephone landline) is about one-third of a monthly salary, at 15 JD per month for unlimited usage. In addition, every minute of Internet access via a landline is billed as a local call, adding to monthly access charges about 1 JD per hour. Another complication is the slow rate of data transfer over ordinary phone lines, which means that it takes large amounts of time to do even simple tasks online, thus costing the user large sums of money. Spam and viruses are other causes for concern regionally, both of which require extra software to manage them and specialized knowledge to contain them as well as other, hidden costs. Economic factors represent a significant incentive to focus an analysis of Middle Eastern Internet culture on the Gulf, where most users reside.

The Internet in Kuwait

An analysis of Kuwait's Internet practices helps us move beyond this general picture to a more careful, focused look at the Internet in context. Kuwait is neither the most wired country in the region (although it was when I began my research for this book in late 1995) nor the least wired; it is not the country in which the knowledge economy has taken the firmest root (the UAE, mostly due to the construction of the Dubai Internet City, a free trade zone for the knowledge economy); it is not the country in which the highest percentage of Internet users is female, nor the lowest; it is not the country with the strictest censorship of the Internet in the region (Saudi Arabia), nor the country with the lowest access prices (Egypt—for years, the Egyptian government offered free access and training to key social sectors such as business, health care, and NGOs); it is not the country with the most Internet cafes (Algeria and Morocco) nor the fewest (Iraq and Palestine); it is not the country with the most ingenious plans for bridging the digital divide or improving opportunities for the poor (Jordan and Egypt). Rather, Kuwait was the first country in the region to give all university students, faculty, and staff free Internet accounts. It was one of the first countries in the region to make regular Internet access commercially available to the public (in 1992).

The real value of the Kuwaiti case is not necessarily what it tells us about Internet connectivity and use in the region, although providing qualitative data on such issues is of value, since there exists so little scholarly analysis on the development and impact of the Internet in the Islamic world based upon fieldwork. Rather, the data yielded by the Kuwaiti case, especially in the area of the social impact of the technology, enable us to see first hand the ways in which local values help to make the Internet and its use a unique expression rather than a patterned response to a technological stimulus. The Internet does not so much determine, but rather is determined by, forces of nation, state, and society.

Internet services were first made publicly available in Kuwait in 1992 as part of the Gulf War reconstruction process. Kuwaitis were regional leaders in terms of developing an Internet culture before the rest of their neighbors. Countries such as the UAE and Bahrain have subsequently surpassed Kuwait in terms of per capita Internet penetration, perhaps because of the slowness with which the government of Kuwait privatized the Internet service provider market and the comparative latency of the information economy in the emirate. In spite of this relative lag, the Internet remains an important part of everyday life in Kuwait. As early as 1997, bumper stickers and business cards displayed company email addresses and home pages. Posters advertised that one could order a pizza (Domino's) or a taxi (Taxi 2000) via e-mail. In Kuwait, people talk about the Internet at work, cafes, home, and school. Training in computers and Internet use is a part of Kuwaiti school curricula from primary school to postgraduate study. Television shows prod people to go online to discuss the latest Internet-based games and to debate how the technology will change the world. Public lectures feature Internet-related topics, including business and society issues in the new media age. Questions about the Internet and the security of Islamic values animate parliamentary debates.

Three main factors help explain the vibrancy of Kuwait's Internet culture. First, Kuwait has one of the highest per capita incomes in the world, estimated at

$23,300 (ppp, 1997) which means that people have money to invest in IT.[8] Second, in Kuwait, new technologies and their acquisition are considered signs of social status, which means that there are common social pressures to be on-line. Third, the government supports a culture of techno-consumerism by enabling the easy flow of high-tech devices into the country. Signs of affluence among the population, such as ownership of the latest techno toys, from computers to convertible Mercedes, are an indicator of an adequate distribution of oil wealth among the population.[9] The importance of technological procurement is seen in the local press and advertising. Regularly, in both English and in Arabic, public media display new technologies for sale in the emirate, sometimes with product reviews. Digital cameras, laptop computers, new software, flat screen televisions, digital phones, and other techno toys are available for purchase in Kuwait as soon as they are released by global manufacturers. This showcasing process is not just "eye candy," as might be the case in other developing countries; rather, each time a new product is reviewed in the press, a list of retailers in Kuwait indicates where such things are available for purchase. A "be the first on your block to have the latest technologies" attitude feeds Kuwaiti techno-culture and subsequently has made Kuwaitis anxious to get online, even if it means paying ten dollars an hour to use a networked computer in an Internet cafe.

One powerful tool for raising public consciousness about the Internet, both its global importance and its local impact, is the newspaper. This view is supported statistically and cross-culturally by a survey conducted by Muhammed S. al-Khulaifi on the influence of the Internet among university students in Saudi Arabia. Al-Khulaifi found that more than half (58%) of those interviewed learned about the Internet from a newspaper, a magazine,

or on television.[10] In Kuwait, newspapers are an important site in which to observe the mediation of Internet
culture. This is true of both the English and the Arabic
language press as illustrated below. Samples of Kuwaiti
Internet culture in public media are drawn from the *Arab
Times, al-Watan, al-Qabas,* and *al-Rai' al-Am,* several of
Kuwait's most widely read dailies.

The English Press and the Internet

During the late 1990s, there were three main ways in
which Internet affairs were reported in the *Arab Times:*
first, as a form of advertising for Gulfnet; second, as a
regular section of local news; and third, through random
coverage of headline news. The paper had a regular classified advertisement called "Internet Pick of the Day,"
which was supported by paid advertising for Gulfnet,
Kuwait's only Internet service provider until 1998. The
daily ad provided a URL with a title or a one-sentence explanation describing the selection. The recommendation
was generally some site that would appeal to a wide audience. For example, the Islamic Presentation Committee
(see chapter 6), a local *da'wa* (Islamic conversion) organization targeting the Kuwaiti expatriate community, was
featured in this column. Also represented was a clearing-
house for all things Indian, linked to the Indian consulate
in San Francisco. Residents of the subcontinent constitute the bulk (53%) of the expatriate labor force in
Kuwait.[11]

The *Arab Times* also carried a regular column called
"Internet News," or "Internet Weekender" (depending on
the day of the week).[12] The "Internet News" section was
offset by a large rectangular box and a shaded title. It always appeared on page 6, and usually took up half of the
page. Internet News was presented as part of the local
news section of the paper. The other half of the page was

devoted to the results of the Kuwaiti stock exchange. Column space within the Internet section was dedicated to
local and regional Internet news along with international
Internet news subjects. Cyberspace subverts traditional
geographic boundaries, and this problematic is characterized by the difficulty in separating local from international news events when discussing the Internet.[13] For
example, the Internet Weekender section combined articles on regional affairs, "Oil Giants Open Net Sites," with
international topics such as "Internet II" being developed
as part of President Clinton's National Information Infrastructure project.[14] Alongside these feature articles were
smaller sections that discussed a particular piece of cyberspace. These entries included PBS's nightly News
Hour, help for teachers online, virus hoaxes, preventing
kids from smoking tobacco, women's networking, a sci-fi
"Web-zine," a NASA newsletter, Web sites for kids' health,
cosmetic surgery, Mexican vacation planning, a sports di-
rectory, dental related web resources, a Chilean newspaper, a site about New York, and learning German online.[15]
At the bottom of the page, the Internet or Internet Weekender section included, in bold print, the following
message:

> For more information on Internet or to access In-
> ternet, please call Gulfnet International on
> 2443800 or 48198733 (24 hours help desk), Souk
> al-Kabir Building, Block A, 7th Floor, Fahad al-
> Salam Street, Safat 13037, Kuwait.[16]

Coverage in the Internet section on page 6 was sponsored
by Gulfnet. Within this section, a direct link was provided
between how to get online and the global reasons one
should. This proved an effective form of advertising for
Gulfnet, when public consciousness of the Internet was
still relatively new. Coverage of Internet affairs and

suggestions of places to surf created a heightened sense of public demand for Internet access, according to Gulfnet officials.[17]

Other ways in which the Internet was featured in the *Arab Times* included headline news, advertisements for home pages, IT trade shows, and Internet-related jobs. Advertisements for Internet-related lectures and human resource development also were regular features. When Internet cafes opened in Kuwait, the *Arab Times* would often give them front page coverage, along with color photos of patrons sipping coffee and surfing the Web.[18] When the Ministry of Communication announced the end of its monopoly on the Internet service provider market, this received second-page coverage.[19] A Kuwaiti businessman's attempts to set up an Open University "using advanced telecommunications systems like satellites, the Internet, and private television and radio stations" made front-page news.[20] Incidentally, this initiative was rejected by the Kuwaiti government, because "it contradicts the country's constitution" (with no explanation of how and why), and thus the businessman, Nasser al-Masri took his dream to Bahrain, "where authorities warmly welcomed him."[21] Also making front-page headlines was an article on the promotion of American alcohol and cigarettes on the Internet and an article about Egyptian terrorists using the Web "to propagate their extremist thinking, their ideology, and to send messages to each other."[22]

The Arabic Press and the Internet

Similar types of coverage of Internet-related subjects appeared in the Kuwaiti Arabic press on a daily basis. The difference between coverage of the Internet in English press and that in the Arabic press is that coverage in the latter was less organized as a subject unto itself. Instead, there was regular coverage of information technology news, of which the Internet was often a subset. For example, *al-Watan* had a regular section that appeared on page 24 called "Computer—a special page for affairs about the computer world and those who work with them." Common features were coverage of regional IT trade shows, articles on Y2K, and articles on new technologies such as "Net-PC," Microsoft® software, and the technical side of broadcasting Saudi-owned satellite television over the Internet.[23] This section generally seemed to be directed more or less toward engineers and other high-tech professionals rather than toward a general audience interested in the Internet.

More closely paralleling the utility of coverage in the *Arab Times* for a general audience, *al-Qabas* had a regular column called "Cruising the Internet," which appeared in the Computer Technology section of the paper. Examples of sites covered in this column were a Web site called "Recipe Finder" (*http://www.homearts.com/waisfrom/recipe.html*)[24] and, on another occasion, a "Discover Islam" site (*httpp://www.discoverIslam.com*).[25] This section usually featured only one site at a time, although with a color picture of the daily selection's home page. The section also tended to be published on Thursdays or Fridays in preparation for the Kuwaiti weekend, which is Friday and Saturday. A recent survey found that 76 percent of all Internet use in the region occurs on the weekend.[26] The sites recommended in this section were usually in English. Many sites on the Internet feature Arabic and now even Arabic based browser software, appropriately named Sinbad, developed by Sakher Software Company (an Egyptian-Kuwaiti joint venture, located in Cairo after the Gulf War). Yet the assumption behind this column seems to be that those who are interested in the Internet will be capable of surfing sites in English.

In addition to coverage of IT-related issues in special

technology sections of the press, Kuwaiti Arabic language newspapers often contained advertisements for new Internet cafes (see Figure 2-3).[27]

Other Internet-related advertising included new software and IT trade shows and seminars, such as those offered by NIIT Institute, whose slogan is "No computers to

Figure 2-3. Advertisement for Kuwaiti Internet Cafe.

know computers in 40 days guaranteed." Editorials occasionally discuss the impact of the information revolution
in Kuwait and the world, such as an editorial by Rida al-
Fili, CEO of Gulf-Ad Com, called "the Technological Revolution."[28] This text became the first in a five-part series
that the author developed over sveral months.

Instead of merely celebrating the Internet and its
promises, Kuwaiti media also considered its potential
threats, from moral corruption and the interruption of
tradition to intellectual decay. One concern raised several
times in the Arabic press was that transitions toward the
use of the computer in daily life would be detrimental
to what little time Kuwaitis, especially youth, spent reading.[29] Occasionally the Kuwaiti press would publish articles and editorials arguing that the Internet supported
activities that were against Islamic values. Other concerns included security risks, both the security of Islamic
values as well as the vulnerabilities caused by Internet
crime—hacking, money laundering, virus propagation,
and networking of criminals. These critical voices illustrate that the Internet is not a tool that is being adopted
blindly by Kuwaitis. New communications technologies
are received, navigated, negotiated, and critiqued, in ways
that demonstrate the panoply of cultural, religious, economic, and political forces that filter a community's response to the Internet age.

A cartoon from *al-Rai'al-Am* summarizes the meta-
morphosis of an Internet culture in Kuwait.[30] It suggests
that familiarity with the use of a computer is an activity
mostly pursued by university students. Before and after
university, a pen, some paper, and a desk are the communication methods of choice. Read from right to left,
this cartoon suggests that we need to look beyond public
narratives if we are to understand how the Internet is mediated by communities in Kuwait. While the press
reaches a wide audience of literate Kuwaitis, this cartoon

Figure 2-4. Table and Chair (read right to left). (1) In primary school; (2) Then to secondary school; (3) Then to university; (4) and afterwards . . . an employee.

implies that the Internet revolution in the Arab world is having the deepest impact on young people, as explored more carefully in chapter 5.

The Kuwaiti Internet user base is estimated to be approximately 8 to 10 percent of the population as a whole (including expatriate labor) and more than 50 percent of the Kuwait University student body. Many residents and citizens of Kuwait have readily available Internet access at work; all students have easy and free access through the University, and many access the Internet regularly at one of the 300 Internet cafes in Kuwait. Since multiple users can be linked to a single Internet protocol address, it is difficult to know what the exact user base of the Internet looks like. The 10 percent Internet penetration figure is based on a 2.5 users per machine estimate established by *PC Middle East* magazine, and thus this is a conservative estimate.[31] The creation of a media and telecommunications free zone in Kuwait (November, 1999) could result in a more vibrant Internet culture as competition in the Internet service provider (ISP) market emerges, and a higher market profile for information technology products and services develops (*http://www. kuwaitfreezone.com*).

The most significant contribution of the Kuwaiti case

to our understanding of the global impact of the Internet revolution is that the meaning and manifestations of the Internet in Kuwait are so contrary to those qualities and expectations laid on the technology by outside observers. We see this in three main areas—political expectations for democratization, economic expectations for privatization, and the emergence of a knowledge economy, and socially, in terms of the emergence of an anything goes, wild frontier of free expression, where power is diffused by the leveling of the discursive playing field, and information is freely shared beyond traditional social boundaries. These expectations and alternative realities are considered more carefully next.

Kuwaiti Internet Practices and Social Scientific Explanations

Internet and Democratization

There was a tendency early on in Internet Studies to link the technology to the promotion of democracy. This is especially true for the literature that considers the Internet and its political meanings in authoritarian countries. The idea is that better access to uncensored information means that citizens are better informed and more likely to make demands on their governments. Empirical evidence, however, suggests that the link between wider access to uncensored media and political activism is not necessarily as strong as assumed. In spite of emerging empirical evidence suggesting a tenuous link between the Internet and automatic democratization, access to and use of the technology are still seen by some as "a barometer for a nation's level of freedom and democracy, its commercial energy, its desire to become part of the increasingly interlinked new world order, and its empowerment therein."[32]

The initial linking of the Internet and democratization is seen in the assumption that "the beauty of information is that it ineluctably democratizes societies," as two prominent Americans have concluded.[33] Some critics have emphasized the Big Brother surveillance capabilities and the invasion of privacy effects of the Internet.[34] But much more common are Barlow-like expectations for the emergence of "a new social space, global and anti-sovereign, within which anybody, anywhere can express to the rest of humanity whatever he or she believes without fear . . . which might undo all the authoritarian powers on earth."[35] As Charles Ess observes, in Western discourse there is an assumption that "in order to construct the new global village as an ethical, social, political, and economic community—all that is needed is to lay the requisite infrastructure of the new communications technologies."[36] As a way to explain the philosophical roots of such optimism, Norman Vig notes that while "relatively few political theorists" have directly addressed the question of the relationship between technology and the process of democratization, "historically, the progress of science and technology has been an important force in liberating peoples from traditional submission and poverty, and thus in establishing the pre-conditions for democratic citizenship."[37] In terms of communications technologies and their effects on democracy, Vig observes that "modern communications technology increases political awareness and creates the possibility of 'direct democracy'."[38] But is creating the conditions for direct democracy enough to engender the end of authoritarianism?

Several scholars of contemporary Middle Eastern studies have probed the relationship between the Internet and democratization in the Islamic world. Their findings are mixed. Some argue that given the tight state controls of the Internet in the region and lack of a mass user base mostly due to cost of access, or lack of infrastructure to support access outside of major cities, Internet use has not had a wide-ranging effect on politics in the region, that authoritarianism persists in spite of better public access to uncensored information, whether Internet or satellite based.[39] Others are more optimistic. For example, Jon Alterman argues that even if access to IT has not really reached critical mass in the region, nonetheless, "change brought on by new technology does seem certain."[40] Some of the long-term changes that we can expect, according to Alterman, include "increasing amounts of information, new ways of interpreting that information, and the rise of new kinds of communities."[41] Observing these same processes in the region, Augustus R. Norton goes a step further in arguing that the Internet is producing "growing civic pluralism in the Muslim world," which will result "in organized demands for equitable treatment by government."[42] What this means in practical terms, according to Norton, is that "the slow retreat of authoritarianism in the Muslim world is under way."[43] Another observer, Dale Eickelman, expounds this view when he observes:

> Whether Arab states like it or not, increasing levels of education, greater ease of travel, and the rise of new communications media are turning the Arab Street into a public sphere in which greater numbers of people, not just a political and economic elite, will have a say in governance and public issues.[44]

Built into all of these narratives is a dose of technological determinism. Such determinism is seen in the idea that the Internet itself is a promoter, or at least an enabler of democracy. We also see determinism linked to the technology when scholars speculate about the inevitability of states in the region relinquishing power and voice to their populations because they are armed with the Internet. In reality, the context of use helps shape the political value of the Internet. There is nothing inevitably democratic imbedded in the Internet. Even though the technology is designed to have power over use distributed among a network of nonhierarchically organized users that is basically free, there is no guarantee that users in authoritarian contexts will choose to say what, when, and to whom they want online.

Part of the problem is an issue of trust, and an issue of reverse capability. Both of these factors make the link between the Internet and democratization less than automatic. The trust issue is that in authoritarian contexts, individuals are unlikely to speak their opinions or organize politically, actions that involve great risk, in places where they do not feel secure. Cyberspace, although it has the guise of security and anonymity for some, in authoritarian contexts it is a place where the audience is unknown and thus not trusted. The second constraint is the Internet's reverse capability. As Gene Rochlin observes, in the Internet age, "the modern firm [like the nation state] puts into place an embedded spider web of control that is as rigorous and demanding as the more traditional and visible hierarchy."[45] In other words, just because the Internet gives the immediate guise of individual empowerment, this should not hide the fact that beyond individual capability lie enhanced institutional capabilities. Perhaps the flexibility of the network gives individuals unprecedented autonomy over their access to information and public voice. But this does not preclude greatly enhanced capabilities for institutions that police and survey individual and collective behaviors online and off-line. The enhanced capabilities of the state in the Internet age make the link between access to the Web and enhanced democratization problematic. As Rochlin observes, "If democracy is defined in terms of power, of the balance between individual autonomy and centralized coordination, the results are at best mixed."[46] This is because the democratization effect is "just a phase."[47] Eventually democratization "gives way to a more stable configuration in which workers and managers [and citizens, in the case of statecraft] find their discretion reduced, their autonomy more constrained rather than less, their knowledge more fragmented, and their work load increased." Moreover, "redistribution [of power] rarely diffuses any of it to those at the bottom of the organizational hierarchy."[48]

Internet and Kuwaiti Politics

One of my goals while in Kuwait was to examine the political effects of the Internet, specifically whether or not the availability of the technology was stimulating processes of democratization. Part of the reason that political use of the Internet to promote democratization has not regularly occurred in Kuwait is that less than 10 percent (2001 est.) of the population has access to the Internet. At the same time, less than 20 percent of the population can vote in national elections. This is because in order to vote in Kuwait, one must be a citizen, male, a property owner, and over age of twenty-one. More than half of the Kuwaiti people are not citizens. At least 57 percent of the society is under age twenty-one. At least half of the people are female. And some Kuwaitis do not own property.

A contributing factor to the depoliticization of the Internet in Kuwait is the fact that there is a public ethic against openly voicing political opinions outside of trusted social collectives, such as mosques, or *diwaniya*/parlor visits (male-female social gatherings within private homes). In Kuwait, the Internet is not viewed as a protected social space but rather as an open, potentially dangerous location where no one really knows anyone else, and one cannot be assured of privacy, immunity, or anonymity. Thus on the Internet, Kuwaitis are in general unlikely to share political opinions outside of the small, nonpublic Listserv™ or private chat room. Public opinion is a bit of an oxymoron in Kuwait. For example, in a survey at Kuwait University in 1997, students were asked to identify their political identity. The most common answer from both males and females was "no opinion." In Kuwait, it is illegal to hold public gatherings without government permission, and political parties are illegal. In 1997, for example, a number of Kuwaiti intellectuals decided to form a new democratic movement. They met once to formalize plans. They tried to meet again to develop their plans further, but the Ministry of the Interior declared the gathering illegal and said that they would have to apply for a license in order to meet publicly in the future. As explored in chapter 3, Kuwaitis who are outspoken are often criticized or brought to trial for their writings. Although not the norm, some are recipients of threats or have even been killed for sharing their opinions publicly. Kuwait is a country where strength and unity of the collective are valued over individual opinion. Thus the Web is not really viewed as a place for activism or distributing political views, outside of the use of the Web by groups trying to propagate Islamic values, as discussed further in chapter 6.

Islamist uses of the Web in Kuwait are said by activists to "not be about politics" but rather "about social reform." This statement reveals several layers of meaning. First, in Kuwait, "being political" specifically means attempting to "share power" with the ones who historically and legitimately, according to the principles of constitutional monarchy, rule, for example, the Sabah family. It also suggests that there is much within social practice that might in other contexts be understood as political behavior, in that Islamists are shifting power to define social practice away from society at large. Another meaning is that Islamists are defining their activities as apolitical in an effort to protect their activism from repression, lest they threaten the state. Thus it is a defense mechanism rather than a statement of fact. Also insulating the state against political activism via the Internet is the use of rents from oil wealth to demobilize the population. A more systematic treatment of the factors shaping political use of the Internet (or lack thereof) in Kuwait is explored in chapter 3.

The Middle Eastern Knowledge Economy

The Internet is viewed as an economic empowerment mechanism or zone by a broad base of Internet researchers and international policy makers. The theme of information haves and have-nots, digital winners and digital losers, those caught within the tides of the digital divide, or those surfing the Web to a whole new realm of economic opportunity is a dominant one in the twenty-first century. A McConnell International report summarizes this collective wisdom when it observes, "World economic growth depends increasingly on information and communications technologies and the abilities of countries and enterprises to collect, process, and use digital information."[49] With the slow growth of Internet access in the Middle East, as suggested by connectivity figures, it is not surprising that economic developments, such as the growth of the knowledge economy, also have been slow in the Arab world. One of the keys to building knowledge economies is establishing information societies. The two go hand in hand. Information society can be defined as "a form of social and economic development where the acquisition, storage, processing, assessment, transmission, and diffusion of information leads to the generation of knowledge and the fulfillment of needs of individuals and firms and thereby plays an important role in economic activity, the generation of wealth, and the quality of life of citizens."[50] Elsewhere I have argued that the four key variables required to build an information society in the Arab world include an IT infrastructure, a knowledge economy, a public culture of discursive openness and formal legal institutions that support the digital age.[51] With this definition, it is clear that the political climate, cultural and social variables, and infrastructure all participate in enabling or disabling the knowledge economy. It also perhaps explains why the knowledge economy has been slow to materialize in the Middle East. Where there is not political will to change cultural attitudes towards the free exchange of information throughout society, growth of the knowledge economy is stunted. Where there is no public demand for IT, and no financial and legal infrastructures to support Internet-based consumer activity, the knowledge economy is slow to grow. Where communications infrastructures are weak and thinly spread, there is an inability to harness a country's knowledge resources beyond the major cities and beyond the upper and middle classes.

Recognizing the potential of the knowledge economy for jump-starting stagnating Arab economies, large amounts of analysis have been generated to facilitate information technology development programs.[52] Seminars, workshops, and conferences are regularly held. Pilot projects are imagined and implemented. Most of this analysis takes place at the level of international organizations and involves key governmental representatives, business people, and leaders of NGOs. Since the knowledge economy in the Arab world is just emerging as a social, political, and economic phenomenon, scholarship on the issue is quite thin.

In an article on IT and economic development in Iran (not an Arab country, but part of the Gulf), Laleh D. Ebrahimian explains the slow growth of the information economy regionally when she observes:

> Over the past two decades, technology has en-riched the social and financial conditions of many countries. The tools have advanced at a rapid pace, transformed every public and private sector, and produced limitless potentials. Yet, the celebration is limited to developed countries.[53]

The case of Jordan illustrates some of the regional constraints on the growth of the knowledge economy. Karla J. Cunningham, in her article on the information revolution in Jordan, explores how the kingdom is at-tempting to harness information technology "as a rapid conduit to economic growth," with mixed results.[54] The basic problem in Jordan and in all developing countries with authoritarian political structures is that these coun-tries "pursue a 'managed' IT strategy that may prompt the regime to limit, in whole or in part, elements of the IT pro-gram if it becomes too costly in terms of resources or un-intended political effects."[55] The problem for the region is that information technology and the Internet "imply an open and open-ended system of communication that many Muslim countries find either economically or politi-cally risky."[56] In Kuwait, constraints on the growth of the knowledge economy are not so much question of the regime's desire to manage the sector to avoid unwanted politicization of information resources as a question of oil wealth dampening the spirit of entrepreneurialism.

Internet-Led Economic Growth in Kuwait

Kuwait is what some local intellectuals call "the last truly socialist state." What this means is that the Kuwaiti state has an elaborate "cradle to grave" network of social services as a way to share wealth generated by the oil economy with its citizens. Some of these benefits include free education (primary, secondary, and higher), paid sabbaticals from government employment every five years, and guaranteed employment in the government sector, which employs nearly 98 percent of the Kuwaiti workforce. Every Kuwaiti is guaranteed free medical care and medicines, a residence, and a monthly family allowance after marriage, and for the birth of each child. No Kuwaiti pays taxes. The government also subsidizes the cost of electricity, water, gasoline, telephone service, and basic foodstuffs.

The result is that in Kuwait there is in general no real notion of socioeconomic competition. Some Kuwaitis complain that a culture of economic entitlement among the population leads to on-the-job apathy, foot dragging, and inefficiency.[57] This noncompetitive and inefficient organizational structure characteristic of Kuwaiti government enterprise, and the increased burden on state coffers caused by commitments to universal employment for citizens, leads some Kuwaitis and many international observers to press for privatization. One observer notes that this is unlikely to occur, because privatization "would lead to the elimination of at least half of the workforce" in the drive to make state-run organizations market driven, profitable, and efficient.[57] Under the present system, private enterprise contributes only in the single digits to the percentage of overall gross national product. Among the few percentage points of the economy located in the private sector are information technology initiatives including Internet cafes, computer sales and training centers, Web page development companies, turnkey services, and software engineering firms, many of which owe their livelihood to large government contracts for computer technology and services. The presence of the Internet in Kuwait has not led to rapid economic growth in the private sector, mostly because the huge government profits from the sale of oil are distributed back to society in such a way as to undermine the energy that might stimulate the growth of the knowledge economy.

The slow growth of the knowledge economy in Kuwait is evident in the local Internet culture. For example, Kuwait was relatively slow (January 2000) to introduce intellectual property legislation (IPR). The presence of such legislation has not helped protect against piracy, which suggests that government enforcement of IPR is lax. Supporting this conclusion is the fact that in 2004, the International Federation of the Phonographic Industry and the Business Software Alliance conducted a study in which they found 60 percent piracy rates for music and 73 percent piracy rates for software in Kuwait.[58] Such figures give Kuwait the highest piracy rate in the Gulf, and the fourteenth worst record world wide.

Kuwait also has lagged regional markets in local software development (perhaps because of all the piracy). For example, it was not until February 2000 that a regional center for software engineering was established in Kuwait with support from the Arab Fund and the UNDP. A similar institution was established in Egypt as early as 1995 with Arab Fund support. Egypt and the UAE also were much more proactive in establishing the Dubai Internet City and the October 6 Media Free Zone, respectively, both of which were up and running while Kuwait was still in the discussion phase about developing a similar project (*http://www.kuwaitfreezone.com*). In Kuwait, the Internet is a "social phenomenon" much more than it is a part of government strategy for economic growth and human capacity building, as explored more completely in chapters 4–6. In Kuwait, the state is too focused on developing the oil and petrol-chemical industries to pay much attention to the information economy. A case in point is the fact that when the government decided to implement an economic stimulus package to jump-start the local

economy in 2000 by pumping $3 billion into new eco-
nomic projects, the lion's share of the investment went to
the oil industry. A small fraction of the money was invested in industrial projects, and no money was invested
in the knowledge economy.

Internet Culture and Individual Empowerment

There is a tendency in Internet Studies to link the
technology to "processes of decentralization, individual
empowerment, resilience, and self-sufficiency."[59] Stated
another way, Rochlin observes that "one of the most persistent arguments for social benefits of the introduction of
computers and computer aided machinery [including the
Internet] revolves around the argument that the personal
computer is an instrument of social democratization."[60]
The expectation that the Internet changes everything, including the distribution of social capital and mobility, is
not met in Kuwait, where strict regimes of hierarchy
linked to tribe, social class, and sect limit the Internet's
ability to empower individual users. The way in which the
structural and cultural givens of Kuwaiti society help
shape individual Internet uses is the subject of the next
chapter. In general, Kuwaitis are chronic communicators;
youths are bored and looking for new "playgrounds"; gender boundaries are strict in real life and easily transgressed in cyberspace; and an increasing number of
Kuwaiti citizens are interested in harnessing the global
power of the Web to spread Islamic awareness. These social factors are predominantly at work in building the
Kuwaiti Internet culture.

Conclusion

Inspired by the centuries' old fascination with the relationship between ways of communicating and ways of
living, the chapters that follow examine how the Internet

is changing the ways in which Kuwaitis communicate, and thus the ways in which they live. The preceding sections explained that Internet use in Kuwait does not necessarily conform to expectations for automatically enhanced democratization, privatization, and individual empowerment. This book argues that Internet culture reflects the context of use. This is especially true in Kuwaiti women's, youths,' and Islamists' Internet practices, as explored more completely in chapters 4–6. Chapter 3 offers a more systematic introduction to Kuwait for those who may be unfamiliar with day-to-day life, economics, and politics in the emirate. It attempts to provide the background necessary to explain why Kuwaiti Internet practices look the way they do; why, in spite of a robust Internet culture, the economy remains undiversified, democracy remains stunted, and social organization persists along tribal, class, and sectarian lines, much the way that it has for more than a century.

CHAPTER 3

Contextualizing the Internet in Kuwait

One of this book's main arguments is that the context of Internet use matters in shaping the meaning of the technology locally and globally. This chapter provides an introduction to Kuwaiti society, politics, and economy as a way to explain some of the peculiarities of Kuwaiti Internet culture. It builds on the cursory treatment of Kuwaiti politics, economics, and society provided in chapter 2 to show why political use of the Internet is constrained, why economic use of the Internet is latent, and why the Internet does not necessarily lead to enhancements of personal freedom and power in Kuwait. Moreover, this chapter examines why gender relations, youth subculture, and religious identity in Kuwait all encourage active use of the Internet. It also provides a basic history of Kuwait, examines factors that shape Kuwaiti communication and media culture, and offers basic reflection on everyday life in Kuwait, inspired by my field diaries.

Kuwait and the Rise of the Al Sabah

The modern state of Kuwait is slightly smaller than New Jersey, occupying an area of 17,820 square kilome-

Source: Map created by Bjorn Anderson, Univ. of Michigan for the author.

Figure 3-1. Map of Kuwait.

ters. Part of the state is bounded by nearly 500 kilome-
ters of Gulf coastline. Seafaring and fishing have contributed to the state's natural resources from its inception. According to Ya'qub Yusuf al-Hijji, "Kuwaitis
elevated the building of sailing craft into a national art
form, a specialty that became the backbone of [the
state's] maritime economy," which included fishing, diving for pearls, and trade.[1] Michael Herb in his text on

Middle Eastern monarchies notes that historically in Kuwait there exists, "an orientation toward the sea," more so than towards the desert.[2] The rise of the oil economy eventually displaced the predominance of the maritime economy, but not the orientation of the country towards the sea, both in diet (large amounts of fish are consumed daily in Kuwaiti households) and in everyday life and leisure (beach houses, beach and sailing clubs, boardwalks, cruising Gulf Road, recreational fishing, jet skiing, etc.).

The modern state of Kuwait rests upon a rich history of inhabitancy, spanning nearly 2,500 years. Archaeological remains from Failaka Island show evidence of an ancient civilization in the region, most likely as a stop on the Dilmun-Mesopotamia trade route. Until the eighteenth century, however, the region known now as Kuwait "was a nameless region on the margins of changing empires, its coasts inhabited by fishermen and its hinterland crossed by nomadic tribes and caravans bound for Constantinople and Rome."[3] This namelessness would change, however, with the settlement of tribal families migrating from the Najd including the Al Sabah, the Al Khalifa, and the Al Jelahima branches of what would become the Bani Utub tribal collective. These families were expelled by the Ottomans from the Najd, because of alleged acts of raiding and piracy.[4] According to the *Encyclopedia of Islam,* the name "Kuwait" has two potential origins. One is "little fort," referring to "a small Portuguese defensive settlement established there in the late 16th century," and the other, less common meaning is "small wells," referring to several freshwater sources near the heart of the city where the town was founded in the eighteenth century.[5]

Kuwait also was known historically by the name "al-Gurain." A town south of Kuwait City still maintains this name. In late twentieth-century history, al-Gurain was the site of an important battle between Kuwaiti resistance forces and Saddam Hussein's occupying army. The Swiss cheese-like carcass of a villa in which resistance forces were hiding and ultimately fighting against Saddam's troops, severely outnumbered and outgunned, remains a museum and memorial. The monument is viewed as testimony to the heroism of Kuwaiti resistance forces during the occupation. Contained within the museum are displays of huge Iraqi artillery shells, the makeshift uniforms resistance fighters wore and narratives commemorating those involved in the fight. On one of the upper floors where one fighter gave his life is the gruesome spray of human flesh. In August 1997 I toured the museum with one of the survivors of the Battle of al-Gurain. The survival strategies used by those few who made it through symbolize the resilience of the Kuwaiti people.

In addition to being bounded by the sea, Kuwait is mostly surrounded by desert. Jacqueline S. Ismael, in her study of Kuwaiti society, notes that the culture of the desert influenced early state formation, social structure and social values.[6] Such desert values are still felt in contemporary Kuwait. Expressions of "desert" culture include the continued rule of the Sabah family, tribalism in politics and society, the persistence of Bedouin dress for both men (*dishdasha*) and women (*abayah, niqab,* and *hijab*). Desert values also are expressed in important public symbols and rituals such as the National Assembly building, which is shaped like a Bedouin tent, and the practice of hospitality, such as offering coffee, tea, and refreshments to guests.

Ismael summarizes the relationship between desert and sea in the early years of settlement in Kuwait when she observes: "Sabah power was based upon the family's relationship to the desert, and consequently their interests were closely allied to the nomadic mode of production, while the economic life of the community was more closely tied to the maritime mode."[7] The Sabah family, under the leadership of Abdallah Al Sabah (1756–1814), had gained recognition as the leading political family in Kuwait, especially after the migrations of the Al Khalifa and Al Jelahima tribes to Bahrain and Qatar. The Sabah family was distinguished by its skills in diplomacy, conflict resolution, and negotiation. Ghanim al-Najjar describes the arrangement between the Al Sabah and the merchants as follows:

> Early Kuwaitis who settled in the area more than two centuries ago chose Sabah the First, the founder of the present ruling family as their ruler. They agreed that he would handle the daily affairs of the society, and that they would support him financially, provided that he consulted with them on major decisions.[8]

During the eighteenth century, real power within Kuwaiti society lay in the hands of the merchant families who dominated and controlled the local economy. Ismael observes: "The real power in the community resided in the financial-commercial class that in effect controlled the development of productive forces. The Sabah house over the period had become financially dependent upon this class and politically subordinate to it."[9] This concentration of wealth and power in the hands of the merchant community would characterize the reigns of the first five Shaikhs, beginning with Shaikh Abdullah I, (1762–1812) and passing to Jabir (1812–1859), Sabah II (1859–1866), Abdullahh II (1866–1892), and Muhammad I (1892–1896). Until Mubarak the Great (1896–1915) staged a coup, killing two of his brothers, and thus seizing political power in 1896, "Kuwait's Sunni merchant notables enjoyed a supremacy over the ruling family."[10]

In 1899, this would change, with the advent of British imperialism in Kuwait (in the form of a protectorate), which strengthened the position and power of the Sabah family. In that same year, Mubarak the Great "imposed a customs levy on the merchants," marking "the decline of merchant political power in Kuwait."[11]

The dialectic between political authority and mercantile power in Kuwait remains a common feature. For example, in 1921 and 1938, merchant hegemony again exerted itself, in the latter case with the support of some disgruntled members of the Sabah family and British interventions to curb monarchic power. The experiment would be short lived, finding expression in the 1921 Consultative Council (lasting only two months) and 1938 People's Legislative Council (lasting six months). Members of the major merchant families populated both of these consultative councils. The first emerged in response to a question of succession in the Sabah family after the death of Selim Al Sabah (1917–1921). The second council emerged as a merchant effort to check political autocracy. With the discovery of oil in 1938, a new informal deal was established by Ahmad Al Sabah (1921–1950) to quiet merchant attempts to intervene in the affairs of state. Under the plan, "in exchange for receiving a sizable portion of oil revenues, the merchants" would renounce "their historical claim to participate in decision making."[12]

During the reigns of Ahmad Al Sabah and Abdullah al-Salem (1950–1965), the merchant families of Kuwait became even richer than they had been in the past, but the price was high. They lost any check they once possessed on the ruling family politically. Oil wealth enabled autocracy and purchased the quiescence of the merchant families in the face of great economic opportunity. The primacy of economic concerns over political aspirations continues in contemporary Kuwait. For example, during an interview with a young member of the merchant class, he observed, "Business, all that Kuwaitis care about is business."[13] Similarly, Ahmad al-Baghdadi, in his effort to explain the demise of several veteran politicians in the 2003 elections observes, "Kuwaitis became bored with politics, opting instead for focusing on immediate economic concerns."[14]

During the reign of Abdullah al-Salem, Kuwait experienced several experiments in political liberalization. Abdullah al-Salem "attempted to lessen the more oppressive aspects of autocracy in Kuwait."[15] For example, "youth associations were allowed to open; local press emerged as the active medium for political debate; elections were organized for important administrative councils; and the foundations of the welfare state were initiated."[16] Abdullah al-Salem accommodated demands, sometimes in direct conflict with other Sabah family members, "coming from the traditional merchant class," and "from political classes that did not even exist prior to the time of his rule."[17] Abdullah al-Salem was renowned for having "created these new classes through extensive programs to redistribute oil wealth."[18] As well, he will always be remembered for having "empowered them politically when he inaugurated constitutional government in Kuwait."[19] Once again demonstrating the tenuousness between liberalization and autocracy, by 1959, "the youth associations and the press were banned, and the elections to the administrative councils annulled."[20] In spite of these setbacks, Abdullah al-Salem created the institutional foundations for representative democracy in Kuwait, including the enactment of a constitution in 1962, the creation of a freely elected parliament (a fifty-member National Assembly), and a formal division of executive, legislative, and judicial powers in Kuwait. The new middle class enhanced public awareness and involvement in politics, whether parliamentary or in associational life. Tensions between rulers and the ruled remained, however, as some members of the ruling family opposed democratization experiments. After the death of Abdullah al-Salem, subsequent emirs have used tactics such as dissolving parliament and/or rigging elections to enhance the government's autonomy.[21] Al-Najjar explained in a recent article what he calls "the precariousness" of Kuwaiti democracy, claiming that fourteen out of the thirty-eight years since the first parliamentary elections have been spent with an absence of parliament either via dissolutions or external challenges, such as the Iraqi occupation.[22]

In addition to manipulations of parliament, the government also keeps a tight reign on civil society. For example, political parties are banned, although "movements" are allowed. In the 1999 elections, for example, six political "groupings" participated, including the Islamic Constitutional Movement, the Kuwaiti Democratic Forum, the Islamic Popular Bloc, the Salifi Movement, the National Islamic Alliance, and the National Democratic Bloc. While these movements can support candidates for parliamentary elections, elected representatives are not allowed, according to parliamentary law, to act as a party once elected. Al-Najjar explains the effect of this situation as follows:

> Under the internal working rules of Parliament, only the government is allowed to speak and be represented as a united bloc; this makes the government the only de facto political party permitted to operate in Parliament. Elected members are not allowed to function collectively, or to have a single spokesman for a group of parliamentarians. As a

result, the government enjoys a stronger position in lobbying for votes on key issues.[23]

Another limitation on political activity in Kuwait is government control of associational life. Haya al-Mughni highlights state control of nongovernmental associations ranging from societies to help the blind to the journalists association to women's groups when she observes the following:

> Kuwait's voluntary associations cannot operate outside the state's institutional framework. In this context, strict administrative and legislative provisions regulate voluntary groups' activities, limiting their ability to pursue their own interests and influence social change in ways likely to conflict with the interests of the state. The state also retains the decision-making power over who ultimately controls an association.[24]

The Law of Public Benefit Societies, according to al-Najjar, "gives the government the full authority to regulate, ban, grant, and license any society in the country."[25] In addition, "the Law of Public Gatherings" restricts "freedom of assembly."[26] Private homes and mosques are exempt from the Law of Public Gatherings, thus lending these two locations an important political role in Kuwaiti democracy. Mary Ann Tétreault has written extensively on the mosque and the home as protected spaces for civic activities in Kuwait. She observes the following:

> Both of these protected spaces support the partial mobilization of political resources and have proven themselves capable of maintaining the capacity of some Kuwaiti citizens to act in the name

of the community even during periods of severe political repression.[27]

In spite of the progressive role that mosque and home
play in Kuwait, Tétreault is critical of their substitution
for real freedom of association. She explains:

> Neither the home nor the mosque is a satisfac-
> tory democratic alternative to political space, pub-
> lic space to which every citizen has ready access
> for the purpose of political action. The home is pri-
> vate, and access is limited to idiosyncratically se-
> lected members of the community. It is too small
> physically and too limited socially to allow the full
> range of citizens' viewpoints on issues to emerge.
> The mosque, though it is public and offers a large
> physical space where people can gather, imposes
> religious controls on expression and action that
> deprive particular groups, such as women and
> non-Muslims, or particular points of view, such as
> secularism, from access to the forum it provides.[28]

The uneasy relationship between state power and social forces demanding more of a say in government has
lively expression in contemporary Kuwait. One is never
quite sure when activism (whether liberal, tribal, Islamist,
female, or mercantile) and the attempts to check political
autocracy will go too far, introducing emergency meas-
ures such as the dissolution of parliament (as occurred in
1976, 1986, and 1999); the careful regulation of associational life (such as the regime's refusal to award licenses
to some voluntary organization, especially in the area of
human rights);[29] and physical violence or arrest (although generally not state sponsored, as explored more
completely later).[30] The uncertainty of cause and effect in
political expression and repression in Kuwait makes the

public sphere a common place for self-censorship and shifting alliances. As discussed in chapter 4, self-censorship and political inhibition shape how the Internet is used or not used in Kuwait.

Oil and Politics in Kuwait

Oil was discovered in Kuwait in 1938 at Burgan field, during the reign of Ahmad Al Sabah (1921–1950), but it was not until 1946 that Kuwait began exporting oil. Jill Crystal summarizes the economic transformation of Kuwait after the discovery of oil when she observes that Kuwait "experienced a radical but apparently smooth transition from pearling to petroleum, poverty to prosperity."[31] The categories "before oil" and "after oil" are common tropes in Kuwaiti historical memory. There is a tendency in the scholarship on Kuwait to explain much of economics, politics, and society in terms of the influence of oil. This tendency led Muhammad Rumayhi, Kuwait's most celebrated intellectual, to write a book in which he argued that Kuwait was not equal to oil.[32] Such a narrow focus, according to Rumayhi, ignores much of the historical and cultural fabric of the region. In spite of a realization that Kuwait is more than the sum of oil's influence in the region, oil is far from benign in shaping everyday life and politics in Kuwait. Oil is directly and indirectly linked to shaping Kuwaiti political and Internet culture. Kuwait owns 10 percent of the world's oil reserves. Reserves are expected to last for the next 100 years at current rates of production, of around 2.4 million bpd.

The discovery of oil and the creation of large rents, which the state accrued and distributed at times to buy loyalty, enhanced the efforts of the Sabah family to rule autocratically. Tétreault, who has published widely on the influence of oil on Kuwait politics and democracy, notes,

"The inauguration of the oil era in Kuwait freed the rulers from their remaining financial dependency on merchant wealth."[33] Oil wealth, however, could not completely neutralize the merchants' and the new middle classes' (a product of the accumulated wealth of the emerging welfare state) demands for checks on political autocracy. She notes that the rents from oil "did not free [the Sabah family] from the merchants' insistence that rulers held their positions not because of royal entitlement but only insofar as they retained the consent of the governed."[34] This dialectic between rent, patronage, competing, and complimentary interests has characterized relations between rulers and the ruled from the beginnings of statehood, whether rents were accrued from maritime pursuits and taxes on trade, or oil. Oil, especially after nationalization of the industry in 1975, clearly concentrated power and influence in the hands of the state. As of 1975, the Kuwaiti government controls 100 percent of the oil industry's profits. The government also owns 97 percent of the land of Kuwait and employs 95 percent of the Kuwaiti population. All of these factors give the government huge power to influence the day-to-day lives of Kuwaiti citizens in terms of the ability to give and to take away material comforts, often in relation to the strength of a citizen's oppositional imagination.[35] When financial livelihoods are at stake, self-censorship can be an auspicious step toward economic empowerment.

The emergence of an export-oriented oil economy in Kuwait intensified the country's links to the global community, building on a rich history of trade and internationalism encouraged by the maritime economy that preceded it. The need for efficient communications with the international community was increased by the export of oil. At the same time, revenues from the sale of oil gave the state the resources with which to build an efficient communications infrastructure. Before 1946 (first commercial-scale exports of oil) and continuing until the late 1950s, communications within and from Kuwait were mostly supported by radio links.[36] A modern telecommunications infrastructure was not operational until 1961, at which time British Telecom set up a 400 line exchange that serviced 200,000 Kuwaitis. In this same year, the Kuwaiti government began to produce and broadcast television programming that rapidly gained popularity in the Gulf. In 1965, Kuwaitis made less than 2,000 international calls. By 1986, that number had grown to 500,000, symbolizing the rapid development of Kuwait's global communication links. In 1969, Kuwait became the second Middle Eastern country to have its own satellite earth station, which by 1980 supported international direct dialing. By 1972, mobile networks became operational. In 1992, after the experiences of the Gulf War, the Kuwaiti government launched a twenty-four-hour satellite broadcasting service with a footprint extending from the Gulf to North Africa.[37] In 1997, the satellite broadcasting footprint of KTV 1 (Arabic) reached all the way to North America in order to spread Kuwaiti-produced news, information, and cultural programming all around the globe. The development of Kuwaiti links to the Internet in 1991-1992 represents another example of Kuwait's highly modern and deeply penetrating web of global communication capabilities. A survey conducted in 1998 by the UAE-based Dabbah Information Technology Group found that Kuwait had the highest density of Internet users per capita of any Islamic society.[38]

In addition to providing the incentive and the means to establish a highly developed communications infrastructure, the oil industry in Kuwait dominates the Kuwaiti economy. Unlike in other developing countries, where there are opportunities for building economic growth through the promotion of information technology initiatives, and through other efforts to build wealth through private-sector investment and development, Kuwait has been slow to adopt such an approach. Current rents give no incentive to diversify. Moreover, oil rents are distributed among Kuwaiti citizens in such a generous fashion that competition and entrepreneurialism within society are reduced. These practices have swelled the public sector and dwarfed the private sector. Guaranteed employment is one of the ways in which oil wealth is distributed back to the population. There are selective incentives not to enter the private sector or to develop a private business, except as a supplement to public-sector employment. For example, work hours are longer in the private sector, pensions are not as generous, and on-the-job demands are more intense.

In other contexts, the information economy is responsible in most cases for building a mass-based Internet culture, both by creating competition in the ISP market, which results in lower costs for Internet service, and by building public awareness and demand for the technology through advertising and stimulating the development of online content. In this sense, the oil economy could be indirectly linked with the lack of a promotion of the Internet as an economic vehicle. At the same time, the redistribution of oil wealth by the state means that those who want Internet access can generally afford it, even if it costs more than in most places in the world. The redistribution of oil wealth has given Kuwait a significant public Internet culture, in spite of the lack of a significant private sector push for the information economy. This curious situation also has raised the social importance of the Internet above and beyond the political and economic incentives to be online, as will be explored more systematically in chapters 4–6.

In addition to giving people in Kuwait the financial wherewithal to afford Internet access, the oil industry also contributes to Kuwait's Internet culture by being a significant developer of online content. For example, the Kuwait Oil Company (KOC) (*http://www.kockw.com*), which was created in 1934 as a British-Kuwaiti partnership, has an elaborate Web site, as does the Kuwait National Petroleum Company (KNPC) *http://www.knpc.com. kw*, est. 1960). Also developing Web content is the Kuwait Petroleum Company (KPC) (*http://www.kpc.com.kw*).

The Iraqi Invasion and Its Aftermath

Just as Kuwaiti national consciousness is significantly shaped by the discovery of oil, Iraq's occupation of Kuwait in 1990 also has had a formative influence on Kuwaiti history and communications strategies. Some of the first Web sites developed in Kuwait were designed to commemorate and raise global awareness about Kuwaiti prisoners of war (POWs) captured during the Iraqi occupation (*http://www.pows.org.kw*). The Internet also was used to help with psychological counseling for those experiencing post-traumatic stress syndrome as a result of the war. Illustrating the importance of the war for Kuwaiti national identity, contemporary history often is divided into a before occupation-after occupation narrative framework.

In terms of Kuwaiti politics, Tétreault observes that in some ways the occupation enhanced processes of democratization. She observes:

> The takeover of a country by a foreign power is hardly recommended as a recipe for expanding freedom and human rights. Yet the outcome of the Iraqi invasion and occupation of Kuwait was to increase the political capital of Kuwaiti opponents of domestic autocracy. The invasion also enlarged the arena where the struggle for democratic reform in

Kuwait was fought. Individuals and groups formerly on the sidelines mobilized to support reform domestically, while foreign constituencies favoring liberalization also expanded. These changes helped to shift the balance between the regime and its opponents in favor of pro-democracy elements.[39]

In addition to stimulating democratic processes, the occupation also had a significant impact on Kuwaiti information and communications strategies. Tétreault argues that the invasion illustrated "the risks of censorship."[40] Another observer notes:

Perversely, the Iraqi invasion of Kuwait has provided positive benefits for Kuwait's telecommunications. . . . The effects of the invasion have allowed it [Kuwait] to plan for a future network that provides a greater degree of stability and resistance to such events. The invasion has also allowed a valuable pause for thought and reflection, a time to evaluate priorities and the needs of Kuwait's population, which have changed as a result of the occupation.[41]

When Kuwait temporarily and unwillingly became part of Iraq (1990), the Kuwaiti ability to communicate with the outside world became a target of Iraqi control and subsequently a source for Kuwaiti resistance. One of the defining characteristics of the occupation was the Iraqi desire to destroy any Kuwaiti ability for selfpreservation. Thus the occupation forces systematically disabled the technological means by which Kuwait had communicated independently, both within and beyond Kuwait. In the service of this goal, the Iraqis destroyed the satellite earth station at Umm al-Arish. The National Museum was torched and its treasures looted. The Iraqis "cut international phone lines and took over or destroyed T.V. and radio stations."[42] Fax machines and radios were rounded up, as were any mobile phones that were found. Archives containing records of Kuwait's history and public discourse were burned or destroyed. The press was taken over and put into the service of defending the occupation and trying to convince Kuwaitis that they were now Iraqi citizens. Public spaces, including streets, buildings, amusement parks, and neighborhoods, were renamed to create an illusion of support for Saddam.

Not surprisingly, in addition to painting over street and neighborhood signs, one of the first acts of the Kuwaiti resistance movement was to reestablish communication links to the government in exile and the allied commanders. John Levins, an Australian author who lived through the occupation of Kuwait, described the importance of these communication links in his memoir:

Within a week of the invasion, the Kuwaiti Government had re-established itself in exile in Taif, Saudi Arabia, and elements of the Resistance in Kuwait began to communicate with it, and with the Kuwaiti embassy in Washington. The most important means of communication was initially a clandestine satellite telephone retrieved from the home of Sheikh Salem Hamoud Salem, in the suburb of Nuzha, on a tip from a Kuwaiti Ministry of Communications technician, Mustafa Qattan. . . . Mustafa was later arrested by the Iraqis and is still missing. During early September, four more US $50,000 satellite phones, arranged by the Kuwait Oil Tanker Company from its offices in Dubai, were smuggled across the Saudi border, concealed in the bodywork and fuel tanks of cars or pick-ups. . . . Each set was more than just a phone with a satellite dish. They could also send faxes and telexes, so were used for written reports, maps, and even photographs. . . . These systems were perhaps the greatest tools of the Resistance, enabling it to maintain direct contact with the outside world.[43]

Not only were satellite phones a vital means of communicating and coordinating resistance activities, but the civilian resistance relied upon other means of high-tech communications tools to define and coordinate opposition to the occupation forces. Levins observes, "Within days Kuwaitis were printing leaflets and newsletters on their home PC's, photocopying them and distributing them by hand or fax."[44] Thus in Kuwait technologies of communication are both symbols of what it means to be Kuwaiti as well as mechanisms through which Kuwaiti identity can be communicated.

In Kuwait, partially because of the experience of being cut off from the world during the Gulf War, ensuring one's access to the latest communications technologies is an important security concern. A departmental chair at Kuwait University illustrates this compulsion to be globally connected. He explains that only a few days after his son's arrival in Australia for college (1997), even before he had become fully settled in his dorm, he called his father and said, "Here dad, take my number, it's for my mobile phone."[45] By having digital, mobile, and globally linked communications networks available at the press of a button, Kuwaitis demonstrate their development, their prosperity, and their power to define themselves in the world. At the same time, having access to digital media is viewed as a way of preventing future occupation and minimizing regional security threats. One reflection of the Kuwaiti desire to project information regarding the country and national identity as a way of reinforcing Kuwaiti independence is the Kuwait Information Office (KIO). The KIO has developed a significant body of web content with which to express the history, politics and culture of Kuwait (*http://www.kuwait-info.com.*) Given the small size of Kuwait's population, and the enormity of the violence perpetrated against anyone who dared resist the occupation, there is not a family in Kuwait who has remained unscarred by the trauma of the war. Its memory remains a defining characteristic of everyday life.

Print Media and Politics

In spite of the Gulf War and the risks of censorship associated with the general public's lack of preparedness for Iraq's hostilities, censorship still continues in Kuwait. Although Kuwait is celebrated as one of only two Gulf countries with an elected parliament and is credited with one of the liveliest and most open print media environments of any of the Gulf States, Kuwait still regularly censors public media. In addition, self-censorship is common. A strict press law limits criticism of the ruling family and its allies, publication of anything considered potentially anti-Islamic, and the distribution of literature considered harmful to public morality. This press law exists in direct tension with the guarantee of freedom of expression by the Kuwaiti constitution. Strict laws governing print media and public discourse discourage freedom of expression, both online and off-line. For example, on May 14, 2003, fifty cyber cafes were shut down by the government as part of an attempt to prevent the circulation of pornography.[46] Moreover, the business licenses of these cafes were revoked until use of the Internet from such cafes could be thoroughly inspected and new legislation regarding Internet usage developed. Similarly, calling into question Kuwait's commitment to freedom of speech and a free press is the government's closing of the al-Jazeera satellite channel's office in Kuwait City. In the first week of November 2002, the Ministry of Information notified al-Jazeera correspondent Saad al-Enezi that it was closing the station's Kuwait office because the station had been deemed "biased."[47] Again evidencing a wavering commitment to freedom of expression and information flow is the fact that books are regularly banned by the Ministry of Information from display at the annual Kuwait National Book Fair. The continued existence of a Ministry of Information and the Ministry's active efforts to police and censor public discourse illustrate the tenuousness of freedom of speech and information in Kuwait.

In spite of censorship, a wide range of media texts is available in Kuwait. Imported print media, including magazines and newspapers, some of which are censored, are readily available in Kuwait, but a large import duty discourages wide readership. In 1997, *Newsweek* began an Arabic language edition, which has some local appeal, but circulation is low compared to local news and cultural magazines such as *al-Mujtamaa* or *al-Arabi*. The most widely read daily newspapers in Kuwait include *al-Anba, al-Qabas, al-Watan, al-Seyassah,* the *Arab Times* (English), *al-Rai al-Am,* and the *Kuwait Times* (English). These papers share a combined circulation of 635,000 as of 1996.[48] Circulation figures for 1993 indicate that *al-Anba* is the most widely read daily, with a readership of 100,000.[49]

Many Kuwaiti newspapers maintain Web sites that distribute Kuwaiti news, information, and analysis to the global community. Interestingly, *al-Anba*, the most widely distributed newspaper in Kuwait, does not have an open Web presence (access as of 2003 is password protected). Similarly, the *Arab Times*, Kuwait's most widely distributed English language daily (50,000), did not have a Web site until late 2003.

Kuwaiti Newspapers Online

al-Rai al-Am	*http://www.alraialam.com*
al-Qabas	*http://www.alqabas.com.kw*
al-Watan	*http://www.alwatan.com.kw*
Al-Seyassah	*http://www.al-seyassah.com*
Kuwait Times	*http://www.kuwait-times.com*
Arab Times	*http://www.arabtimesonline.com/ arabtimes*
al-Anba	*http://www.alanba.com.kw*

Journalists and authors who overstep the boundaries of Kuwait's press laws often face persecution, harassment, and in some cases, they have even suffered bodily injury or imprisonment for what they have written. In 2001, a Kuwaiti lieutenant colonel and police officer gunned down and murdered Hedayah Sultan al-Salem, editor in chief of *al-Majlis* magazine. Hedaya had written an article in which she stated that the traditional female dance of the Awazimi tribe, to which the assassin belonged, was sexually suggestive. Ziab Khaled al-Azmi, the assassin, claimed during his trial that he was angered by the article, published in *al-Majlis* in July 2000, as it was said to insult the Awazimi tribe, and thus he shot her. Hedayah was sixty-five years old when she was gunned down. She was one of Kuwait's first female journalists and was very active in the struggle for Kuwaiti female suffrage and full political rights. On February 4, 2003, Khaled al-Azmi was sentenced to death. An appeal was made, but on June 8, 2003, the death sentence was upheld by the appeals court.

In 2000, two authors, Layla al-Othman and Alia

al-Shuayib, along with their publisher were brought to trial for their writings. Dr. al-Shuayib was immediately dismissed from her position at Kuwait University but later reinstated. The Kuwaiti court found both al-Shuayib and al-Othman guilty but commuted their sentences to a monetary fine. Also tried was Professor Ahmed al-Baghdadi, who was said to have published an article in the Kuwait University student newspaper that said that "the prophet failed in Mecca to win converts." Any act of failure attributed to the prophet is blasphemous, according to the court's ruling. Al-Baghdadi was sentenced to one month in prison and fined, but poor health resulted in the commuting of his sentence to the thirteen days he spent in jail and a fine. These cases suggest that journalists and authors face uncertainties that can limit freedom of speech. The vague nature of the press law, as well as the possibility that aggrieved citizens may take matters into their own hands in opposing media discourse, means that self-censorship is common. In spite of government censorship of the media, Kuwait still has one of the liveliest political cultures in the region. We turn now to the shifting alliances and debates of the Kuwaiti political scene.

Identity and Alliance in Kuwaiti Politics

Kuwaiti politics is characterized by a rich variety of cleavages and coalitions that on the surface conceal a highly complex and rapidly shifting body of power relationships. In contemporary Kuwait, one of the key divisions in politics is the line between individuals who want to promote liberalism and Western values (or locally defined versions of democratization and modernity) and individuals who are interested in reviving or enhancing

Kuwaiti tribal and desert traditions. Another division is the line separating liberals and tribal factions from those who want to enhance the place of Islam in everyday life, the Islamists. These three divisions—liberal, tribal, and Islamist—are the predominant ideological divisions within Kuwaiti politics, in addition to sectarian and class differences and general patterns of regime loyalty (e.g., pro-government politicians vs. opposition forces) that cut across these ideological lines. Often there are alliances between members of these ideological traditions, in parliamentary politics and beyond. Tribalists and Islamists often are in agreement on promoting more conservative laws governing, for example, relations between genders in Kuwait, such as the separation of males and females at university. Shamlan al-Essa, a professor of political science at Kuwait University, explains cooperation between tribal and Islamist forces in a more strategic way when he observes that "in their rise to political prominence, the Sunni Islamists have in turn used the Bedouin [triabalists]—still the most traditional elements of Kuwaiti society—as their main recruiting ground."[50]

Liberals and Islamists, although for different reasons, sometimes find themselves sharing opposition to tribal politics when opposing what is viewed as the promotion of backwardness, raw tribal self-interest, and/or ethics of the Jahiliyya (age of ignorance, pre-Islamic nomadic identity). Both liberal and Islamist movements often share a platform that is modernist in the sense of promoting gender equality, urban values, knowledge, science, and capitalism. We see modernist Islamist values expressed in the uses of the Internet to promote Islamic sciences and economics, as examined in chapter 6.

There are elements within the Islamist movement, however, that are antimodernist and share more with Bedouin/tribal factions, as in the case of Islamist MP

Walid al-Tubatabai, who opposes giving women the right to vote or to run for public office, and has argued in Parliament for the need to censor the Internet.

Islam and Politics

According to Shafiq al-Ghabra, former director of the Kuwait Information Office, and president of the American University in Kuwait, Islamists are "the only organized mass based political force in the country."[51] In the wake of September 11, and increasingly controversial U.S. foreign policy decisions in the Middle East, liberal forces in the Arab world, Kuwait included, are finding it more and more difficult to curry favor among a public convinced that the United States is building a new cold war, this time with Muslims as the opponent. The polarization of the world into the West against Islam resonates with Islamist definitions of world politics and seriously undermines the Muslim liberal attempt to successfully promote a politics of compromise that has Western-style democracy, liberalism, and secularism as its goal.

The reticence of Kuwaiti liberals to define a political and social vision that is both authentic and modern, but distinguishable from Western paradigms, adds fuel to the flames of Islamist opposition and success. Islamists have an easier deal to sell, a return to practices that historically have made Muslims world leaders. Liberals, on the other hand are trying to sell a deal that often looks like mimicry of a foreign culture and, increasingly, a culture that is defined as being hostile to Islam. We see the weakness of the liberal platform and the strength of Islamists in Kuwait clearly expressed in the 2003 election results, whereby the liberals dropped from six seats in Parliament to only three seats, while Islamists mostly held their ground with seventeen seats (in 1999 they had eighteen seats). As the only organized, mass-based political and social movement in Kuwait, it is perhaps not surprising that Islamists are well represented in Kuwaiti Parliament, and in Kuwaiti cyberspace.

Since the 1980s Islamist politics can be divided into three main affiliations: the Social Reform Society, whose ideological orientation is toward the mainstream Muslim Brotherhood (Ikhwan); the Salafi (ancestral) movement whose institutional foundation in Kuwait is the al-Turath (heritage) organization; and third, the Shi'a movement, known in Kuwait as the "Jamiyyat al-Thaqafah al-Ijtimayyah," or the "Social Cultural Society," which draws strength and inspiration from the Iranian Revolution.[52] Two areas where Islamists have had a deep impact on Kuwaiti society are in the Ministry of Education and the Ministry of Information. In terms of education, al-Ghabra observes that Islamists have since the 1980s "laid the groundwork for a more conservative school curriculum."[53] Moreover, Islamist influence at the Ministry of Information has resulted in more conservative broadcasting and increased censorship. Both education and the media are important tools with which to shape public culture.

Islamist control in these two areas in part explains the increasingly conservative orientation of Kuwaiti public life. Another factor that explains Islamist power and influence is, according to al-Ghabra, "the amount of funding, equipment, and staffing available to the Islamic groups," which is "much greater than what had been available in the past to any political group in the Middle East."[54] Islamist movements also are represented by alliances in Parliament.[55] In 1992, the first elections after the liberation of Kuwait, Islamists were the dominant force in parliamentary politics, which, according to Tetreault, "guaranteed the prominence of religion in National Assembly politics."[56] In the 1996 elections, nearly half of all seats went to Islamist candidates whose power in Parliament is often enhanced by alliances with tribal and pro-government forces. In contrast, only eight Parliamentary seats went to liberal and secular forces (1996), and liberal representation would continue to decline (1999, six seats; 2003, three seats).[57]

Some liberal Kuwaitis argue that the government encouraged the growth of the Islamist movement in Kuwait, hoping to develop a foil to growing liberal demands on the state. Now, however, Islamists have grown in number and in strength. They are lobbying the state even more effectively than the liberals, and as a result, the character of Kuwaiti society is thus being reshaped to conform with the perceived values of the Shariah. Islamists are increasingly making demands on the government, as represented by criticism from the Salafi movement. Secretary general of the movement, Hakim al-Mutayri, called for "implementing a true democratic system in Kuwait, similar to the one practiced by civilized nations."[58] His criticism of the present system was that there is an "inability of the Kuwaiti people to effectively choose their government, and an inability to hold it accountable when it fails."[59] Islamist MP al-Tabatabai, also of the Salafi movement, criticized the government for reserving "key ministerial posts for members of the ruling family" (members of the Sabah family control the Ministries of Foreign Affairs, the Interior, Defense, and Finance; they also control the National Bank, in addition to controlling the position of prime minister and emir).[60] He also criticized the government for failing to adapt Kuwaiti law to be completely compatible with the Shariah. These outward criticisms of the government by members of the Islamist camp illustrate a shared platform between Islamist and liberal oppositional politics. In addition, such open opposition to the state illustrates that the Kuwaiti government may have miscalculated its ability to co-opt Islamist forces as a foil to liberal opposition movements. This coalescence between Islamist and pro-democracy opposition also explains the state's efforts to introduce pro-government tribal elements into Parliament as a foil to both Islamist and Liberal opposition.

One of the ways the government of Kuwait helped build an Islamist movement, with the hope of foiling the increasing power of liberal politicians, was to support the establishment of Islamist institutions. In addition to supporting NGOs and community organizations, the government has helped Islamists build three high profile institutions, the Kuwait Finance House (see chapter 6), the College of Shariah at Kuwait University, and the Committee for the Implementation of the Shariah.

College of Shariah and Islamic Studies

The College of Shariah was established in 1981 and is located on the Kaifan campus of Kuwait University. The college has its own library and its own cafeterias, both of which are strictly segregated by gender. Nearly all women on campus wear Islamic dress, including the black *a'bayya*. Most men have beards and wear short *dish-dasha* (a typically long, flowing robe-type garment worn by Kuwaiti men as a form of national dress), both of which are symbols of Islamic conservatism in Kuwait. Classes also are segregated, with women studying separately from men. Interestingly, however, male professors still teach female students, often in small seminars. Most of the women attending the College of Shariah and Islamic Studies plan to enter the field of education or library and information sciences. Many of the librarians at the College of Shariah are also students at the school. I was invited to attend classes with these women, and I found the classes very well organized, open to discussion and debate, and relatively balanced. The principles of research and the careful pursuit of knowledge were applied to classical sources of Islamic *fiqh* (jurisprudence). Students had a natural enthusiasm for their studies and received guidance from professors often trained at al-Azhar or in Saudi Arabia. Those trained at al-Azhar tended to be more liberal, whereas professors trained in Saudi Arabia tended toward the conservative Wahhabi interpretation of Islam. The College of Shariah has a mosque, as do the other campuses of Kuwait University. Attendance at daily prayer and the Friday sermon at the Kaifan campus, however, tends to be greater than at other campuses. The librarians at the College of Shariah in 1997 began developing their own CD-ROM databases of classical Islamic texts, *hadith, fiqh,* and Quran. Moreover, one professor at the college developed his own online journal. Students from all over the world come to study Islamic law at Kuwait's College of Shariah. Many come on full scholarships, giving the school an international as well as a religious character. In the words of Shafiq al-Ghabra, the College of Shariah is viewed by Kuwaiti liberals as "a school of traditional religious indoctrination; these are the schools that produce Islamic activists. Most of the instruction is provided by professors who tend to be either fundamentalist or orthodox."[61]

The College of Shariah and Islamic Studies serves the growing needs of government schools and mosques in Kuwait. With the activism of Islamists in the Ministry of Education, greater demand for teachers qualified in Islamic studies has emerged. Students in every grade level take daily religion courses in every government school in Kuwait. Moreover, the rise of other Islamist institutions, such as the Committee for the Implementation of the Shariah and the Kuwait Finance House as well as their increasing presence in the Kuwaiti National Assembly and the Ministry of Information, has increased the demand for scholars trained in the Shariah, those who can guide organizations and individuals in interpretations of correct Muslim responses to the challenges and opportunities of the twenty-first century. The student body of the College of Shariah also has grown, from only 160 students the college in 1992 to more than 1,200 students attending in 2001. The college has its own Web site that can be accessed at *http://shariah.kuniv.edu.kw/*.

Committee for the Implementation of the Shariah

In 1992, following the liberation of Kuwait, Sheikh Jaber al-Ahmed al-Sabah established the Committee for the Implementation of the Shariah. The goal of the committee was to create a community of scholars that could review Kuwaiti laws in order to propose modifications that would make all laws of state and society compatible with the Shariah. The committee also has been active in reviewing and reforming Kuwait's school curriculum, public broadcasting, and other aspects of community life so that public culture is in line with the Shariah. On the tenth anniversary of the committee's establishment, Sheikh Khalid al-Mathkour, head of the committee, said in an interview that the most important achievements of the committee include integrating "the study of the Holy Quran in the school curriculum at all grade and year levels," amendments to Kuwait's civil code, and the establishment of an institution called "Kuwaiti Establishment for Information Improvements," which will find "substitutes for all programs that do not match with the Islamic teachings."[62]

The committee is housed in a huge white villa (provided by the emir) in one of Kuwait's wealthiest neighborhoods. The villa includes Dr. al-Mathkour's office as well as offices for his staff, meeting rooms, and an elaborate library, with books on Islam and Islamic law collected from all over the world. A special team of engineers develops on-line resources and databases for the committee. This team is also housed at the villa. Dr. al-Mathkour is an al-Azhar-trained sheikh. He is well respected throughout the Islamic world for his moderate interpretations of Islam. To liberals in Kuwait, however, his views are anti-quated. For example, he issued a *fatwa* that said that the Barbie doll is un-Islamic. He is quoted as ruling, "She's no innocent doll. She's a mature woman who wears accessories and revealing clothes and has a boyfriend."[63] At the same time, Dr. al-Mathkour has chosen a middle ground between Islam and technology. He observes that if "Muslims are careful about how they use modern technology," there is no incompatibility between Islam "and the modern life."[64] For example, he argues that "mobile phones and the Internet are good" unless "used for flirting," at which point "they become evil."[65]

Kuwaiti liberals view the Committee on the Implementation of the Shariah as a sign of government favoritism and failure to heed the long-term, unintended consequences of supporting Islamist objectives. The large amount of government funds awarded to the organization and the activism and impact of growing religious conservatism are interpreted by liberals as taking Kuwaitis backward rather than forward. In the words of Ahmed al-Bishara, head of Kuwait's National Democratic Movement, "either they take us back to the 7th century or we take them to the 21st century."[66] Liberals fear that once Islamists conquer Kuwaiti public culture they will take on the government next, and that soon Kuwait will look like Saudi Arabia. Liberals would prefer that Kuwait look more like Dubai or Singapore.

Through institutions such as the College of Shariah, the Committee for the Implementation of the Shariah, the Kuwait Finance House (see chapter 6), and Parliament, Islamist forces in Kuwait have been able to slowly, yet steadily, redefine Kuwaiti public culture in more religiously conservative ways; in the words of Khalid al-

Mathkour, "to promote the good teachings of Islam in the country."[67] Islamists and liberals are on opposite sides of the social and political fence. One Islamist, Abdal Razzaq al-Shayji, assistant dean of the College of Shariah at Kuwait University, summarizes the relationship between Islamists and liberals when he observes: "There is a battle here over the character of society—we want to Islamicize it, and the liberals want to secularize it. The liberals want to bring American values to Kuwaiti society, and this contradicts our norms." The 2003 election results suggest that most Kuwaitis agree with Shaiji.[68] According to Nassir al-Sane, an MP for the Islamic Constitutional Movement, the main areas in which Islamists are working to promote observance of the Shariah include "banking, education, IT, health, and the family."[69] Chapter 6 will continue the discussion of Kuwaiti Islamists' attempts to transform public space and cyberspace, especially in terms of IT, banking, and health.

Gender and Politics

In addition to questions about religion and politics, much of Kuwaiti social and political life is intersected by questions about gender. Kuwait is a country where women face daily challenges because of their gender. The inability of Parliament to achieve a majority vote on an emiri decree granting women the right to vote (1999) symbolizes the challenges women face in Kuwait. A Kuwaiti cardiologist, Dr. Farida al-Habib, chief of cardiology at Kuwait Armed Forces Hospital, epitomizes the gender contradictions that characterize Kuwaiti women's lives. She notes that every day I "enter the small veins and arteries in the hearts of men" to unclog blockages, "yet they block me from voting."[70] Other challenges include acts of physical violence against "un-Islamically dressed" women. In one case, a young woman was attacked and her arm broken by an Islamist organization known as "The Desert Flogging Group" for being "inappropriately" attired.[71] While such acts of violence are uncommon in Kuwait, they symbolize the ways in which Islamic conservatism is increasingly at work on women's bodies—disciplining, dressing, and directing females about how to act in public space and determining what is not allowed. Lulwa al-Mulla, secretary general of the Women's Social and Cultural Society, Kuwait's oldest women's organization, notes in response to the shrinking places for women's self-expression, "I am very, very threatened by what they [Islamists] are doing. I do not see a bright future."[72]

In the press, in private conversations, and online, as will be explored more completely in chapter 4, the boundaries of patriarchy are tested by female social critique. Part of this discursive resistance is stimulated by the fact that women are denied many of the benefits provided to Kuwaiti men, such as government-supported housing. A Kuwaiti woman is only guaranteed government housing through marriage to a Kuwaiti male. If she marries a non-Kuwaiti male or remains single, she is unable to get government housing benefits. If she marries a non-Kuwaiti and has children, they are denied Kuwaiti nationality (and a whole host of government benefits) because nationality is determined by the husband in a marital relationship. For example, if a Kuwaiti male marries a non-Kuwaiti, he inherits all the same government benefits, and so do his children. This arrangement puts pressure on female citizens of Kuwait to get married, and to marry a Kuwaiti. A Web site for the Embassy of Kuwait summarizes the problem of women in Kuwait when it observes that within the family and within society, "males carry more prestige than females."[73]

One professional, single mother (by divorce) whom I

interviewed stated that even renting an apartment is a problem for unmarried women. In Kuwait, landlords do not want to establish a rental contract in a woman's name. Thus this woman, who has a Ph.D. and is a mother, had to ask her younger brother to sign for her. Her brother also had to sign the contract to sponsor a maid to care for her son, to get a telephone, and to buy a car. Single women who lack male relatives are severely encumbered. Some Kuwaiti women prefer to hold together even abusive marital relationships in order to avoid the social stigma and built-in difficulties of being single and female in Kuwait. One Kuwaiti woman who has studied gender politics in Kuwait summarizes the situation in this way:

> Women are still being persecuted for committing so-called 'moral crimes.' They have no legal protection against any form of abuse within marriage and no citizenship rights similar to those of Kuwaiti men, and face constant discrimination at work. Given the uncompromising stance of male society, it is clear that the challenge facing Kuwaiti women is daunting, and changes will be slow to achieve.[74]

Women in Kuwait, however, are much better off than women in many other Gulf societies. In Kuwait, women can drive. Women are subject to compulsory public education from grades 1–12. Women attend university and are commonly awarded government scholarships to study abroad. Women make up the majority of students in the colleges of medicine and science, as well as the majority in the college of education at Kuwait University. Thirty percent of the Kuwaiti workforce is female (including expatriate labor), and of this female work force, two-thirds are married. In the public sector, which employees 95 percent of Kuwaiti citizens who work, half of the employees are women. The government guarantees employment to all citizens, both male and female, if they want to work. Marriage and/or gender do not necessarily preclude women from working, although there is an active public discourse trying to drive women into marriage, stating that "marriage is one of the signs and proofs of Allah in the universe."[75] This same Islamist discourse encourages married women to stay at home, stating that "the real place of the woman is in her home . . . raising children."[76]

Despite conservative public discourse, women in Kuwait (including married ones) are represented in all of the professions, including medical, legal, academic, and business. Women are well represented in print and electronic journalism as well. The Journalists' Association in 1997 elected a woman, Fatima Hussain, to its board of directors. At the time, she also was editor in chief of a prominent women's magazine, *al-Samra*. Women also are a major part of the support staff that keeps complex government bureaucracies in Kuwait running, and the sense is that the society could not function without 50 percent of the small Kuwaiti society working (e.g., if all or most women stayed home). In 1996, Kuwait modified its labor laws to be more sensitive to women's needs and to meet international standards. This governmental action can be interpreted as a further entrenchment of women's presence in Kuwaiti public life. While women are employed throughout Kuwaiti society, most of the leadership and upper-level management roles are reserved for men in both the private and public sectors. In the words of one observer, "Whereas the West has a glass ceiling, in Kuwait it's concrete."[77]

While Islamists try to drive women back into the home (or into fields "compatible with their nature," such as education and nursing), liberal women propose counterarguments that stress that men need to share responsibilities at home, because women increasingly share responsibilities to provide.[78] At present, it is quite common in Kuwait to see fathers out with their children at the store, at the movie theater, and having lunch, although it is unusual to hear of men cooking or cleaning at home. In an interview with a middle-age Kuwaiti woman it was noted that "younger generations of Kuwaiti men were more open to sharing responsibilities at home with their sisters, mothers, and wives. Men forty and above would never be caught doing 'women's work.'[79] When asked about the causes of change in the younger generations attitudes toward women and work, this woman observed that the younger generation grew up with satellite TV, was more likely to travel and to study abroad, and thus was accustomed to different gender roles than its fathers and grandfathers. It is possible that the Internet also is playing a role in such changes, as suggested by the narratives in chapter 4.

Haya al-Mughni has analyzed how patriarchy and the state affect women's organizations. She notes, in terms of women's associations:

> The current control of women's groups by elite and upper-class women is by no means accidental. Linked by common interests, the state and women from privileged classes worked to exclude women from lower classes to hold leadership positions and/or create associations. Moreover, the patriarchal social structures regulating Kuwait's civil society restrict women's participation within state-controlled groups. This leaves Kuwaiti women with little autonomy to organize themselves to pursue strategic gender interests.[80]

Al-Mughni's analysis illustrates both the state's tacit control of associational life and also the possibility for partnerships between the state and merchant-class women. The discouragement of a mass-based women's movement in Kuwait is reflected in the depoliticized nature of women's Internet use in Kuwait. Some women who are online feel inhibited to speak freely; other more politically minded women such as Badria al-Awadi expressed in an interview that she felt that the Internet did not offer Kuwaiti women much opportunity to address those issues which most needed addressing, such as helping illiterate women understand their rights in marriage. She expressed that if the Internet had any political importance for Kuwaiti women, it was in terms of giving them access to the ideas and initiatives of the international women's movement.[81] As explored in chapters 4 and 5, this perspective may be changing, as some young women in Kuwait find the Internet an effective tool with which to oppose growing Islamic conservatism.

Youth Subculture and Everyday Life in Kuwait

Approximately 57 percent of the Kuwaiti population is under age twenty-five. This means that youths constitute one of the biggest segments of the Kuwaiti population. In spite of being the dominant group within the society, Kuwaiti youths often complain that their country has little to offer them in the way of entertainment. This is especially the case for teenagers. Bored teenagers can influence public safety, especially road safety. In Kuwait, the most common cause of childhood death is auto accidents. In one case during my fieldwork, an adolescent member of the Kuwaiti ruling family, along with a peer from a leading merchant family, borrowed a family car and, without a driver's license, took it out on an evening joy ride. The car crashed, and both boys were killed. Adolescent suicide also is a problem in Kuwait, as is drug abuse.

Some Kuwaitis blame the traumatic effects of the Iraqi occupation for the challenges Kuwaiti youth face in the twenty-first century.

A recent article on motorcycles and Kuwaiti youth subculture reinforces the notions that the challenges youth face are linked to the effects of the Iraqi occupation, and also to the constraints of a religiously conservative society that allows few public outlets for youths' self-expression. The article notes: "Thrill seeking and a taste for the fast life are the flip side of what Kuwaiti psychologists have said is a state of almost permanent nervousness created by Iraq's 1990 invasion and the fear it could happen again."[82] One of the bikers interviewed, Mohammed al-Rifa'i, explained his participation in such a dangerous hobby, "tearing up and down Kuwait's seaside cornice, dodging police cars, weaving in and out of traffic and . . . wowing girls" in terms of Kuwait's limited offerings of alternatives for youths.[83] He stated: "Entertainment is mostly for families. There aren't any discos or parties here."[84] Out of a gang of twenty-five bikers, four have lost friends in crashes in a single year, again highlighting the public safety risks of teenage recklessness.

In response to the challenges youths face in Kuwait, Islamists in the Ministry of Awqaf have launched a campaign to remind youths of the importance and duty of religious observance.[85] A public information campaign established by the Ministry of Awqaf in October 2003 encouraged Kuwaiti youths not to miss their prayers. The campaign came in response to a study by the Ministry of Awqaf that revealed that less than one in four Kuwaiti youths prays regularly. The goal of the campaign is "to embed a set of religious values and concepts to encourage the youths to perform their prayers regularly."[86]

In the Kuwaiti youth subculture, as in Kuwaiti society at large, a set of polar opposites between forces of increasingly persistent Islamic conservatism and forces of increasing liberalization competes. One sphere in which to observe this identity struggle among youths is dress. For example, some youths choose to wear the latest Western fashions, including baseball caps, baggy jeans, and Western sports jerseys for men and tight, low-cut jeans, midriff shirts and other European designer fashions for women. For Kuwaiti males to choose Western dress over national dress, the dishdasha, is sometimes a form of rebellion or resistance. For example, two opposition politicians, made public statements by wearing business suits instead of the *dishdasha* in Parliament. On the other side of the political and social spectrum, Islamists and Muhajiba women send their own public signals via dress. Male youths who are Islamically conservative usually have beards and often wear *dishdashas* 20 centimeters shorter than standard to communicate their affiliation with the Islamist camp. Young women, who are religiously conservative, tend to wear a *hijab* (head scarf), an *'abayya* (long, black wrap worn over clothing), and sometimes *niqab* (full-face covering), all of which pronounce their allegiance to conservative religious practice.

One measure of growing Islamist influence on the Kuwaiti youth subculture is the issue of veiling for young women. Part of the importance of veiling as a symbol of public identity comes from the fact that in the 1950s and 1960s the unveiling of women was an important sign of liberal change in Kuwaiti public life. Haya al-Mughni explains that unveiling became associated with *nahda*, or "progress," and veiling with *rajiya*, or "backwardness."[87] Until the 1960s in Kuwait, al-Mughni observes, "Women were not allowed to go to the *suq*, or to the tailor to be fitted for Western-style clothes, or even to be driven around the town in a private car without being veiled."[88] Among Arab nationalists, the liberation of women was seen as part and parcel of building an emancipated nation, one that was strong, modern, and capable of being scientifically advanced. In the words of one observer, "We want Kuwaiti girls to rebel against traditions, first and foremost against the black *abbaya,* which inspires depression and grief."[89] Shafiq al-Ghabra observes that as a result of this rebellion against tradition, during the 1960s and 1970s, "few women wore the Islamic *hijab* (which permits only the hands and face to show). Restrictions on the mixing of the sexes were not rigidly observed, and regulations inhibiting women's participation in sports and many kinds of work were loosening."[90]

Given the importance of unveiling as a sign of modernity and national advancement in Kuwait, the Islamist movement's promotion of veiling as a sign of authentic and correct Muslim conduct is met by some liberal Kuwaitis with a sense of hostility and resentment, as if in the words of one liberal female activist, "they want to take us back in history rather than forward."[91] In the face of the liberal critique that wearing the *hijab* is like turning back the hands of time and erasing progress, Islamists argue that the veil is a symbol of female piety and is not in any way incompatible with modern life and scientific advancement. To veil is to fulfill one's duty as a modest woman, and to do otherwise is to violate the ideals established for proper social relations determined by the Quran and the Sunna. In response, liberals argue that the Islamic teachings on veiling are not transparent, and that veiling or not veiling is a matter of personal choice and should not be imposed upon a woman by the state or social movements. The fact that more and more women are choosing to veil in Kuwait is a particularly sensitive issue for Kuwaiti liberals, many of whom have active memories of the struggle to remove the veil from public life and associate modern dress with the promotion of women's advancement in society and politics. In response to growing conservatism in Kuwaiti dress, many liberal women in Kuwait offered the following assessment: "Just wait, in five years, Kuwait will look just like Iran. We'll all be veiled head to toe."[92]

Yusif al-Ibrahim, former dean of the School of Administrative Science at Kuwait University in 1997, and now Minister of Finance, explained in an interview that the trend toward Islamic conservatism can easily be grasped in terms of looking at changing dress habits among students at Kuwait University.[93] He suggested that if one compares yearbook pages from the year he graduated from Kuwait University in the early 1960s to those from 1993 and beyond, the contrasts are stark. Islamic dress is the dominant mode today, but was relatively uncommon in the 1960s. In the 1960s, famous Kuwaiti feminist Fatima Hussain and her cadre were burning their veils. In the twenty-first century, Kuwaiti women are wearing the *hijab* in droves and are proud to do so for reasons of modesty and piety. If one looks at photographs of public events, graduation ceremonies, for example, in the local press, and one counts veiled women versus unveiled women, the results confirm al-Ibrahim's observations.

This table suggests that young Kuwaiti women are nearly five times as likely to veil as to not veil in public life. Liberal women have used the veil, or lack thereof, as a symbol of modernity, freedom of expression, and liberation. In the face of increasingly conservative dress, they find a wider social conflict with their own deeply held personal values. Some of the liberal women I interviewed ex-

Table 3-1. Photographs of Female Graduation Events in Kuwait City.

Newspaper	Date/Page	Veiled	Unveiled
Arab Times	March 20–21, 1997, p. 2	26	5
Arab Times	June 7, 1997, p. 6	6	1
Al-Watan	April 4, 1997, p. 6	34	6

plained that they are dismayed by the pressures they often experience from their children, both male and female, who encourage them to wear the veil, so that they "fit in better and don't embarrass the family." I explained to these women that I had increasingly experienced pressures from conservative women with whom I had been conducting interviews to wear a veil. They said in response, "Don't you dare give in. This is the epitome of what we are fighting for, our rights."

Within these struggles to define acceptable youth conduct in everyday life, communications technologies help both sides. For example, cell phones enable religiously conservative females to interact with males in ways that would not be acceptable face-to-face. One case in point is the religiously conservative Internet cafe user who dialed the manager's mobile phone number to ask him to turn down the air conditioning. They were separated by less than ten feet of space, yet norms discouraged the young woman from directly asking for what she needed. The same cafe manager noted that the young, religiously conservative woman who designed his company web site was able to conduct business at night with him, interacting freely in cyberspace, whereas under normal circumstances, such practices would have been culturally unacceptable. In this same Internet cafe, a professor from the engineering department at Kuwait University, who is an Islamist, regularly brings groups of curious students to learn how to surf the Net in ways compatible with Islam.

At the same time, communications technologies aid more liberal youths in Kuwait to lead lives more in line with Western values, in spite of the general restrictiveness of Kuwaiti culture. For example, the spread of mobile phones has enabled young men and women to call each other at home, without their parents' awareness. This practice has facilitated more open contact between unmarried youths. Moreover, mobile phones are used to arrange meetings among youths of the opposite sex. For example, it is common for young men to drive by the cars of young women, especially on Gulf Road, and throw phones into the car. Another youth will call the mobile number with a second phone, and lines of communication are established, often leading to meetings between the groups of youths at the fast food restaurants or supermarkets that line Gulf Road. This is known as "highway courting."[94] For some, the Internet also seems to be supporting the equivalent of an online dating service, which enables transgressions of gender lines outside and beyond strictures that prevent comfortable and easy interactions among youths in real life.

Regardless of the adolescent struggles of Kuwaiti youth, there can be no doubt that Kuwait is a society that celebrates both family life and the importance of children. Kuwaiti society is designed around the family as a unit of social, economic, and political organization, from the Sabah ruling family to the merchant and professional families to tribal and expatriate families. Entertainment is structured with families in mind. Amusement parks, picnic areas, and restaurants with playgrounds, beaches, and boardwalks all cater to the needs of the Middle Eastern family. Symbolic of the importance of family life are the many centrally placed playgrounds in Kuwait. The faculty club at Kuwait University has the most ornate and appealing structures for climbing, swinging, and sliding. Play equipment is brightly colored, protected from the sun, and equipped with fans to circulate the air during the warm summer months. The Sultan Center in Shuweikh, a gourmet's shopping center delight, has another well-laid-out playground, beside which exists a small restaurant that sells delicious *shawarma* sandwiches. Tables and umbrellas are near the playground so parents and older siblings can relax while younger children play. Of course, as discussed earlier, older children sometimes find that the focus on the family limits places for independent self-exploration within Kuwaiti society. In this sense, the Internet for some provides a ticket to social experimentation, even if not completely free of contextual constraints, as explored more completely in chapter 5.

Conclusion

Key features of Kuwaiti society—politics, history, economics, identity, and culture—as explored in this chapter help explain why Internet practices in Kuwait are not necessarily similar to Internet practices elsewhere. With such factors as the oil economy, the autocratic tendencies of the Sabah family, gender constraints, constraints on youths, and increasing Islamic conservatism, it becomes clear how and why the Internet is generally not used in Kuwait to promote large-scale privatization and transitions to the knowledge economy, democratization, and increasing freedom of expression for the masses. It also is relatively easy to understand why and how women, youths, and Islamists are the dominant social groups attracted to the powers of the Web. We will now turn to these more specific examples of Kuwaiti cyber culture.

CHAPTER 4

___

Women, Gender, and the Internet in Kuwait

Introduction

Over the past few years, many high-profile women have drawn attention to the problem of women's low access to the Internet in the Middle East and worldwide. In 1995, Dale Spender, in her book *Nattering on the Net*, observed the following:

> Despite the belief of some individuals, the computer is not a toy; it is the site of wealth, power, and influence, now and in the future. Women—and indigenous people, and those with few resources—cannot afford to be marginalized or excluded from this new medium.[1]

At the World Bank's Global Knowledge '97 meeting, Lourdes Arizpe, assistant director-general for culture at the United Nations Educational, Scientific, and Cultural Organization (UNESCO), observed that women in the Arab World and beyond, if given access to the Internet, can use these technologies to present their autonomous voices in the service of their own culturally diverse and regionally specific forms of liberation. Several years later, Queen Rania of Jordan, in a speech to the 2nd Arab Women's Summit, observed, that "It is important for Arab women to make use of the latest technology, particularly the Internet, to reshape their lives."[2]

Najat Rochdi, president of Morocco's Internet Society, explains that the Internet can be used to "expand women's leadership skills." When describing what exactly the Internet might do for women, she observes that what needs to be changed is "the culture of machismo in Muslim countries," whereby "women are only valued as mothers and caretakers. We are still not valued as clever and accomplished people on our own accord, outside of the influence and tutelage of a father or a husband. This relates to a lack of trust in women and our leadership potential in both civic and political spheres."[3]

In January 2003, i3me, a Dubai-based information consulting firm, organized in Beirut the first of what it hopes will be a series of conferences called "Women on the Web." Head of marketing and sales for i3me, Shirin Motamed observes, "To be able to access information, or simply use e-mail is empowering" (http://www3.estart. com/arab/women/www.html). Motamed explained, "Keeping the conference just for women should encourage their participation . . . if it was for both sexes, many women wouldn't come because of social taboos." When asked about the goal of the meeting, Motamed explained, "We want to start a 'web' family by connecting women from different countries through the Internet. Imagine all these women from Iran, Lebanon, Jordan, or Egypt being able to exchange information and communicate. . . . The group could bring a new face to Middle Eastern feminism. Or on a smaller scale, simply help keep families which are so often spread out, in touch."[4]

In spite of the clear acknowledgment of the importance of the Internet to women in the Arab world, it is estimated that only 4 to 6 percent of all Middle Eastern Internet users are female.[5] When compared to female Internet usage in the rest of the world, it becomes clear that women in the Arab world are potentially in the deepest recesses of the digital divide.

To contribute to the growing concerns about women's access to and use of the Internet in the Arab world, this chapter records several examples of women in Kuwait narrating their relationship to the Internet, and it uses these narratives as a window to those aspects of culture and power that regulate women's daily lives. This chapter also examines some reasons access to the Internet does not necessarily determine the result of use. In this sense, it discusses why the Internet's presence in Kuwait will not automatically revolutionize women's relationships to formal institutionalized power. Theorists in North America and Europe are fond of arguing that "for those in possession of information technology, power, influence, privileged status, and domination are further enhanced and assured."[6] Women's lives in the Gulf and in Kuwait in particular, however, suggest that advancement, even for those with access to the Internet, will continue to be con-

Table 4-1.

Internet Usage by Women by Region and Percentage (2002)	
Arab States	6% of all users
China	37% of all users
European Union	25% of all users
Japan	18% of all users
Latin America	38% of all users
Russia	19% of all users
South Africa	17% of all users
United States	50% of all users

Source: Women's Learning Partnership: Technology Facts and Figures

textualized in everyday forms of struggle and victory that aim to carve out spaces for freedom in the face of deeply entrenched hegemonies of patriarchy.

Kuwaiti Networks and Women's Voices

One of the best ways to understand how the Internet is affecting women's lives in Kuwait is through oral testimony of the participants. The voices of women who are active Internet users reveal important characteristics regarding the cultural frameworks that regulate both women's lives and their "networks" in Kuwait. Through these examples we obtain glimpses of the promise and problems of new communications technologies for women in the Arabian Gulf. These particular narratives were selected because of the women's differences in age, status, nationality, profession, and perspective to provide a representative cross section of the larger community of Kuwaiti women that uses the Internet on a daily basis.

Nassima

Nassima[7] runs the learning resource center at a private school in Kuwait. She is middle aged, a Kuwaiti citizen, and a self-taught computer technology expert. Every day she introduces new Internet users to the tool's power, yet her words remind us of the ways in which the Internet can reinforce boundaries between genders, if not deployed in ways compatible with women's lives in Kuwait. I visited her at the school the last week before classes let out for summer recess in June 1997. We spent an hour together talking about the Internet and education and gender issues, as well as viewing some of the educational materials provided by the school to guide Internet use. We laughed together when we viewed the "bookmarks" that students maintained on the center's Netscape-based Web browser. The bookmarks suggested that the Internet connection at the school was a tool for male pleasure rather than female gender education and resistance. Nassima observes:

> Girls don't use the Internet unless required to for a class. Boys come after school and use it for pleasure. They go to sites with cars, sports, pop music. The only time girls got actively involved was when they were using the Internet for horoscope information. Girls have a different attitude towards technology than boys. Boys are comfortable with it and like to play with it. Girls are not comfortable with it and would much rather giggle together and talk. Boys teach other boys how to use the Internet and how fun/useful it can be. Boys don't teach girls for obvious reasons, and few girls, if any, are highly skilled in the technology and able to teach others. Thus girls don't learn to be comfortable with the technology in a nonthreatening way. Girls are expected to go home after school. Boys are able to come after school to the center to play with the Internet. Once a boy tried to access a site on explosives. A message appeared on the screen, "You are forbidden to go here." Forbidden by whom to this day we still do not know. The Ministry of Information censors our Internet guides. Here, look at this Web magazine. The cover advertised a story that discussed love on the Internet, and looked at how boys and girls are developing relationships through the IRC [Internet Relay Chat]. The Ministry censored this.

Nassima's narrative, on the surface, seems to reinforce stereotypes about girls and their relationships with technology and science. She notes that girls are unlikely to use the Internet in their free time and are unlikely to teach other girls how to use it. These conclusions contrast highly with the fact that when young Kuwaiti women reach university, large numbers of them choose to major in fields such as computer engineering and pre-med, and they perform well. At Kuwait University, for example, there are more women than men in science and medicine programs, and female students in the sciences continuously outperform male students in terms of grades and test scores. In 1997, for example, Kuwait University raised the test scores required for female entrance into the engineering department, while it lowered standards for male students, because women were scoring better on entrance exams than men, and the school wanted to balance enrollments. The reason given was that administrators wanted to avoid placement problems for graduates, stating that the market could not support large influxes of female labor.

Nassima's own story of being a self-taught Internet expert, being very comfortable with Internet technology, and being able to teach girls how to use it challenges her own explanation. When her narrative is viewed in light of the testimony of other women interviewed for this book, it becomes clear that Internet use at this particular school is influenced by the administration and architecture of the lab and thus is not necessarily representative of young Kuwaiti women as a whole. For example, at the Learning Resource Center, the only time for free use of the Internet is after school, when girls are most likely not able to "stay after." Families in Kuwait tend to keep track of their girls in a way that they do not track boys. Girls have a clearly defined place within the home and the extended family network to which boys are not as strictly subjected. Perhaps if Internet free-play hours were available for girls during the regular school times, they would be more apt to play with the technology. Fieldwork in Kuwait suggests that when homes have an Internet connection, girls are just as apt as boys to use the tool. Both boys and girls tend to gather with friends at home or at Internet cafes to surf the Net.

The physical layout of the Learning Resource Center also limits female use. For example, only one computer is available with an Internet connection. If boys are using this machine, girls are unlikely to play along with them. Perhaps if the Learning Resource center had separate hours for girls during regular school hours, girls would be more likely to learn the technology. If more than one computer had an Internet connection, say one for girls and one for boys, or a few for each gender, then this also would enhance girls' Internet access. If a boy is using a computer, a girl is unlikely to ask him to move to give her a chance, thus having only one supposedly gender-neutral computer makes this computer a typically male domain.

Layla

Layla is a prominent Kuwaiti woman in her late fifties. She is originally from England and is Kuwaiti by marriage. She has been subjected to the pressures of the family matriarchs reserved for "foreign imports" into the bloodlines. These experiences have made her more vocal about women's issues in Kuwait, and her voice has not been one to advocate their need for liberation. On the contrary, Layla observes, "Women in Kuwait aren't in need of any more power. It is in their nature to want to control everyone and everything around them. Arab women are strong, and many are mean. I'm afraid of them." I heard versions of these observations from many Kuwaiti men. One man told me that I "don't understand the distribution of power in Kuwaiti society if I think

women needed any more power. They already control society." Another told me that Kuwaiti women "have tradi-
tionally managed social and financial day-to-day life in
Kuwait because of the heritage of pearl diving. Men went
away for months at a time to dive. Women were left behind to take care of the community." When I told these
stories to another Kuwaiti women's rights activist, she
dismissed them, stating,

> That's the problem we're trying to respond to.
> Men, even liberal ones, view us as already powerful
> enough. But when one asks them to define
> women's power, it's a very domesticated view, stay-
> ing home taking care of the house while the men
> go out and dive for pearls. These historic images
> are slowing women's advancement even today.
> They bind women's as much as men's imagina-
> tions.

As Layla's testimony reveals, part of the problem, as defined by women's rights activists, is the need to convince
Kuwaiti women to demand more than domestic forms of
power.

Layla's testimony encourages us to look beyond the
rhetoric of patriarchy which, on the surface, conceals
women's power in Kuwait. Underneath these more public,
institutional expressions of patriarchy lies a private world
that women rule. From this perspective, women control
the private lives of men, and they run the country from
behind closed doors. Harriet Beecher Stowe called this
kind of power "pink and white tyranny."[8] Cartoons in the
Arab world commonly poke fun at this relationship by
representing a large, strong, powerful woman hovering
over a cowering male. The caption commonly has the
woman making some demand of the man, and the man
rarely is allowed to disagree. But what about the women

who want more than to control power from behind the scenes? Does the Internet offer Kuwaiti women any alternatives for redefining their power?

My visit with Layla occurred in one of the reception rooms of her home. I began by explaining my interests in the ways that new communications technologies such as the Internet might hold meaning for Kuwaiti women. In response to a question asking her whether or not she sees the Internet as a positive force for women's empowerment, she said:

> Well I don't think women in Kuwait will use the Internet for positive social change. They are lazy and would rather talk about superficial things like makeup and fashion. Women are also inhibited in what they say publicly to protect their reputations.

In part, the critical voice behind this narrative attacks Kuwaiti culture and the values it places on women's beauty. What is surprising, or perhaps not, is that many of these standards are defined and maintained by other women. After all, it is mostly women who see other women unveiled. Mothers look for potential mates for their sons at wedding receptions and parties where large groups of young women appear unveiled. In Kuwait, women often offer unsolicited advice on each other's appearance, in ways that may be considered rude in other contexts. For example, several times I was told by other women that I was "getting fat, and better watch out because my husband would be unhappy." I was told by some women that I should buy more expensive clothes or wear more makeup because the impression someone makes through dress and appearance is very important in this part of the world where men run things. Men also help preserve the notion that part of a woman's value is in her appearance. For example, one colleague, before a meeting he set up for me with a powerful member of the ruling family (patriarchy at work), asked me if I had to wear my glasses and told me to dress in an appealing way, and not to wear my "ethnic jewelry." He said that he was giving me advice as a friend; that Arab men enjoyed admiration from beautiful women. One woman told me that she would much rather have male bosses than female ones, because women were "back-stabbing, jealous, and unstable. Men were predictable and easily manipulated by women's ways." Such sentiments help bolster patriarchy, and its subversion, by using female sensuality (or insecurity) to build and interrupt women's social standing. During my research on the Internet in Egypt (2000), one of the women I interviewed stressed that the Internet held value for her because it lessened the importance of appearance in social interaction. In Kuwait, however, appearance continues to be a significant variable in how a woman's value is determined publicly and privately, in spite of the Internet.

During our conversation, I told Layla about my experiences with Nassima, the head of the Learning Resource Center, and discussed girls' relationships with the Internet in the private school I visited. I shared with Layla my theory about Internet "free time" and its inaccessibility for girls at the school. I ask her about the family constraints placed on children and about her interpretation of girls' freedom of movement and association, and whether or not she thought girls are more encumbered than boys in Kuwait. She observes:

> I suppose that's right. I guess girls are subject to different expectations than boys. The truth is that there is pressure on all family members to be at home together for the midday meal. Afterwards everyone takes a nap. If boys wanted to skip this meal, then their families would probably overlook

their absence and would explain the situation in terms of boys needing solidarity with their peers. Girls, however, need to express their first loyalty to the family and are expected to be at home, protected and safe.

"And to help with the dishes," I muse, remembering my own childhood. She laughs:

No, Kuwaiti women and girls do not help at home. They have maids to take care of domestic responsibilities, to watch the children. It's just the pressure to be home together to which girls would be held more strictly to than boys.

Layla is one of the only women I have found whose organization has a home page on the Web. I asked her about it and she told me that she "helped design it." I wanted to ask her to what degree she helped, but I, did not dare, lest I lose my almost empty category "Kuwaiti women developing content for the Web." The last time I probed a women's organization about its Web site, the leader told me that her husband designed and maintained it. Layla's response that "she helped design the home page" is probably more representative of a woman with the financial resources to hire another to do the labor-intensive part of the job and to make it look more professional rather than indicating her own technological apprehensions.

This interpretation is supported by the rest of our conversation, which, revealed an intimate awareness of how the Internet works and careful consideration of the implications of the Internet for Kuwait. In this regard, Layla observes:

I'm worried about what the Internet will do to Kuwait. First of all, this society is not prone to read. The Internet, like satellite TV and video games before it, further encourages Kuwaiti youths to avoid reading books. I'm not sure that they will use the Internet for serious research, as they lack the skills to search for information that is not easily accessible through personal association. I'm most worried about how it is changing youths' attitudes towards sex. We're seeing it in schools now. Students are more comfortable interacting across gender lines than ever before. Young people are experimenting with their sexuality in ways not common to this conservative society. Adults would rather close their eyes to these changes and pretend that traditions live on. These are circumstances in which youths will not get the information they need to be safe in their experimentation, that is, until there is a problem which is out in the open and publicly acknowledged; one which cannot be concealed. This same process happened with the drug epidemic we now face in Kuwait.

Layla's narrative helps reinforce the images, presented in a more sympathetic voice, by Su'ad and Badriya, as analyzed later. Su'ad and Badriya are both part of this new youth subculture, which Layla characterizes as "interacting across gender lines ... experimenting with their sexuality in ways not common to this conservative society." To a degree, the Internet is helping to support this culture of openness toward new gender values (discussed further in chapter 5), but public sanctions on such openness still remain the norm; and it is in light of increasingly conservative Islamic values (as discussed in chapter 6) that right and wrong are judged. In spite of the fact that Islamic values seem to slow the rapid transformation of gender relations in Kuwait, experimentation, as described by Layla and the young people in-

terviewed for chapter 5, cannot help but leave such relations more carefully scrutinized, challenged, and perhaps transformed.

Su'ad

Su'ad is an electrical engineering student at Kuwait
University. She is twenty years old. She began using the
Internet in college and admits that at one point she became addicted to it and had to quit cold turkey for several
months until she got her use under control. Now she limits herself to once a day. Her narrative is important, because it provides a contrasting image to that constructed
by Nassima. She explains:

> I use the Internet every day. I come to the lab
> and use IRC. My little sister uses it too. She's been
> using the Internet since she was six. She's eight
> now. People hack around all the time. Here in
> Kuwait, many people use other people's accounts
> to surf the Net. They break into places where
> they're not authorized to go. I hack. The Internet
> cafes are full of users. Have you ever been? I'll take
> you there and teach you to use IRC. In Kuwait, if
> you give people freedom, they will misuse it. This is
> why the Internet is dangerous. Kuwait channel one
> on IRC is all about sex. I prefer Kuwait channel two
> instead. I meet interesting people there. One man
> who is Kuwaiti but is studying now in London has
> been pursuing me on IRC. He sent me his picture
> as an uploaded file through IRC. I got it, and I
> started laughing. He looks just like my father. I
> could never marry him. I want someone very hand-
> some. I told him this, and he said to give him a sec-
> ond chance because the picture was not really a
> good one . . . I have a friend who is getting married

to someone she met on the Internet. They only "chatted" for four months, and now they're going to spend the rest of their lives together. I think she is stupid. It's possible to lie on the Internet. How does she know that he is really as good as he says he is online? One has to be careful. . . . One time I was "chatting" with another engineer from Saudi Arabia. He kept asking, are you a man or a woman. Finally, I answered, "I'm a woman, is this important?" He said, "Yes, I refuse to talk to you."

Su'ad's narrative provides an image of a woman at ease with technical environments. Her words are representative of the many young women at Kuwait University who are specializing in the sciences and are serious about their advancement within Kuwaiti society. Many young Kuwaiti women major in the sciences, because their chances of employment in the medical and scientific fields are high.[9] Like many of these young women, Su'ad wears a *hijab* (head scarf), but this seems an almost irrelevant detail. Many women veil because of the anonymity it provides them in public places. Being veiled can enable women to meet and speak with members of the opposite sex, a practice that could be difficult if everyone knew who they were. Su'ad explains, when asked about veiling, that she is respectful of her Islamic values, yet she observes that she was raised in a liberal environment, by parents who were educated in the United States. Thus, she feels comfortable crossing strict gender lines on IRC and in real life. For example, once when Su'ad took me to an Internet cafe, she said that someone she met on IRC would be meeting us there. "A guy?" I asked. "Yes," she said, "a guy, don't worry, I know he's OK. I asked my friends about him, and they said he was worth meeting." I was surprised by her boldness, because of the rules restricting interaction between males and females, especially in public places. When we met this man at the cafe, we found out that he was in his twenties and was an electrical engineer who worked as a troubleshooter for Kuwait Airlines. He joined us at a table, and so did the cafe owner, also a male in his late thirties. The four of us sat and talked about how the Internet was changing Kuwaiti society. Our conversation was perhaps symbolic of the broader changes taking place in Kuwait. The computer brought us together, the computer cafe provided the context, and new freedoms of transgender interaction in cyberspace made us all comfortable sitting and conversing face-to-face in this new contested space.

Su'ad's willingness to teach me how to use IRC challenges Nassima's observation that girls do not teach others how to use the Internet. Her willingness to help guide me through IRC's special linguistic codes reveals how the educational process works. In return, I have also been asked by women in the labs at the university, or in Internet cafes, to help them when they are just learning IRC. I'm not an expert in chatting, but often there is a male who is, sitting nearby. Local hegemonies do not enable women to ask men for help. The more women who are comfortable with new communications technologies, the more examples there are to follow, and the more potential teachers there are who can help women go online. While for some users, going online in Kuwait means entering a whole new world where men and women learn to interact with each other in nonthreatening and previously inadvisable ways, these interactions continue to be conditioned by the local codes of a conservative Islamic environment, symbolized by the Saudi man's refusal to even "chat" with Su'ad on IRC. The advent of the telephone, the cell phone, the shopping mall, and the automobile has not rendered benign the effects of conservative Islamic culture on Kuwaiti lives. Observers should not expect the introduction of a tool such as the Internet to do so either.

Men and women mingle on IRC but are segregated in Internet cafes. To mingle in cyberspace is safe, but to do so at the mall or on the street is not. Thus in general, new technologies are adaptable to local environments, and usage conforms so as not to cause open and offensive violations of local cultural codes. In Kuwait, and in other places such as Singapore, in terms of Internet use, equilibrium exists between "permitting room for creative expression and maintaining society's moral standards."[10]

Badriya

Badriya is a computer science major at Kuwait University. She is nineteen and originally from Iran. I interviewed her in the computer lab at Kuwait University. Badriya is conservative and wears the *hijab*, but she, like Su'ad, is liberal and outspoken. Her narrative emphasizes the way in which the Internet is changing women's status, at least in cyberspace. Her words also remind us of the contextual hegemonies that prevent the use of the Internet for open and active gender and political resistance.

> I use the Internet daily. I use it mostly for entertainment purposes, when I get bored, which is often. There's not much for young people to do to relax in Kuwait, [so] the Internet fills this social gap. One of the things I like most about the Internet is that it allows girls to speak with authority, whereas in real life, men constantly tell women that men are superior and that women shouldn't speak. Still, I don't think that the Internet will support active struggles for women's rights. Politics are dangerous here. People are afraid to speak. If you're an important person, or a person with connections within Kuwaiti society, you can say whatever you want. If you're a small person without

public importance, you cannot; you lack protection. Most women lack protection. Even some important people are afraid to speak in Kuwait. People can take what you say the wrong way and use it against you, so most people just maintain a low profile to protect their reputations.

Badriya's narrative emphasizes once again the way in which the Internet has the power to change a woman's voice by deemphasizing the gender of the speaker. In spite of these changes in a woman's voice online, such narratives remain contextualized in, for example, a legal system where under certain circumstances a woman's testimony only counts for half of a man's testimony. Kuwait is a society where a century ago houses were designed so that women's voices would not be heard by visitors. One Kuwaiti historian elaborates by observing that "it was considered 'aib (shameful) to let women's voices be heard."[11] These attitudes toward women's voices still have an impact on the female voice today. Unwritten rules govern when a woman should speak publicly. Given the emphasis on marriage as a social given, young women are careful about how they behave verbally in public. Outspokenness may be interpreted negatively by potential suitors and by other women. Thus in courses at Kuwait University, where I occasionally guest lectured, I observed that women were highly unlikely to speak in class, even though they usually ended up with better grades, which meant that their silence was not about lack of understanding. The main difference between Western women and Kuwaiti women is that Kuwaiti women, even so-called liberated ones, are unlikely to act or speak in a way disapproved of by their husbands and male relatives. Women in the West, however, are more free to act (to speak) as they please (at least this is how they are perceived by Kuwaiti women). So although gender neutrality provided by some transactions in cyberspace enables some men and women to learn about each other's lives, these cyber liberations remain constrained by history and local culture on the ground.

Badriya's narrative also reminds us of the power constraints on people's voices in general. If one is not a person with *wasta* (an Arabic term for "connections,"), then one is not protected from the potential harms of speaking out. Women tend automatically to have less wasta than men. Even those who are from prominent families are very careful about what they say. Throughout my research in Kuwait, women would utter, "Don't quote me on this," and "Off the record of course." Most women who were cautious about being quoted had some story to tell, that revealed why they were concerned. One woman, who has a Ph.D. and comes from one of the leading families in Kuwait, explained that a man she knew spoke out against an Islamist member of Parliament and as a punishment lost his summer teaching opportunity. Thus she notes, "we need to be careful."

Another Kuwaiti woman who holds a Ph.D. noted that she once read an interview, conducted by *The Observer,* with two Kuwaiti women studying in London. The interview touched on sensitive issues such as virginity, women's honor, marriage, and women's rights. My friend was taken aback by these two women's boldness. She had read the published interview in *The Observer* while she was studying abroad. While talking with a friend some years later on the issue of women's voices, my friend recounted her amazement at these two women's boldness in *The Observer* interview. Her friend said, "Yeah, you remember what happened to the one who was not protected by *wasta!*" My friend said no and explained that she was out of the country when this article was originally published. Her friend told her that as soon as word of the published interview reached Kuwait, it stirred quite a scandal in the local community, and both girls' reputations were tarnished. The mother of the less "protected" of the two students feared that this blemish on her daughter's reputation would prohibit her from being "suitably married." Thus the mother flew immediately to London and demanded that her daughter be married to one of the sons in the second outspoken student's family. Evidently the student with "protection" (by her status) encouraged the other one to participate in the interview. As repayment for her irresponsibility and insensitivity, the mother demanded that her daughter be married to one of the family's eligible sons as compensation. Several months later, the less protected young woman was married.

Another person I interviewed explained that during the first Gulf War (1990–1991), she and a friend gave an interview to a Western newspaper describing women's activities against Saddam's occupation forces in Kuwait. They gave the interview based on conditions of anonymity, but the reporter revealed so much personal information about the two women that "everyone" in Kuwait knew they had spoken publicly about issues many Kuwaitis wanted to keep private. Ever since this experience, she has kept a low profile.

These stories are examples of the kinds of narratives (perhaps urban legends?) that are told to reinforce boundaries between public and private discourse in Kuwait. The attempted murder on a Kuwaiti colleague of mine and her husband as they drove home from their chalet one evening in the spring of 1997 reveals the costs of being outspoken even for the very well connected. Both she and her husband have rich public records of organizing for women's rights (the wife, an outspoken women's activist) and fighting corruption in the misuse of public funds (the husband, a veteran MP). The sense was that this murder attempt was directed at the husband for his campaigns in Parliament to oust those government

employees who were looting the nest egg of future generations through the embezzlement of public funds. Evidently he got too close to publicly embarrassing some
very important people, and they tried to permanently remove him from office. This event took place in the middle
of my fieldwork and was very disturbing, as it revealed
the degree to which some individuals will go to keep certain kinds of information private. Kuwait shows signs of
democracy, such as a relatively free press and an elected
and outspoken Parliament, yet simmering beneath the
surface is a public fear of overextending one's freedom of
speech. These narratives of "punishment" for speaking
too frankly reveal the range of subtle to extreme methods
by which Kuwaiti society polices public behavior in random ways. Kuwait is not a police state; it is, rather, a very
small community where everyone is related in some way,
and where strength of the community is valued over the
right of the individual to speak. If corruption exists,
which it does in every society, it is sometimes best not to
discuss it openly, so that from the outside, appearances
suggest that all is well. The status quo is valued over
change. Anyone who publicly advocates for change has to
be prepared for a whole range of possible consequences.

Because voices are constrained by social sanctions
against speaking out, most individuals in Kuwait do not
feel comfortable using the Internet to publish information
that could be used against them in the "social courts" po-
liced by their neighbors, relatives, employers, and friends.
Cyberspace is an extension of the realms of social practice and power relations in which users are embedded. At
times, voice is liberated from gender restrictions, such as
from within the cyber relations enabled by IRC. But voice
is historically subjected to constraints based upon publicly enforced notions of right and wrong in public discourse. The advent of new forums for communication

does not automatically liberate communicators from the cultural vestiges that make every region particular; that hold society together. In Kuwait, this means that women are not likely to organize and speak out against their husbands, brothers, sons, fathers, or bosses, to publicly embarrass their patriarchs. It is more likely that in the privacy of an office with the door shut, or in the living room during the hours that many husbands are at work, these voices will be lifted. These voices once raised are likely to cut to the quick of an issue, to express a well-reasoned, culturally seasoned opinion of women's lives in the conservative Gulf. These voices, if uttered by liberal women, might stress that men in the Gulf are simply afraid of women and their power and are unlikely to yield to women's desires if overt and confrontational demands are made. Some liberal women stress that women's struggles for liberation in the Arab world require subtlety and compromise rather than all-out revolution. Seduction and charm are the best tools for carving out spaces for women's freedom. If the voices come from Muslim conservative women, they are likely to point to the kinds of oppression women in the West face, from which women in the Islamic world are free, such as the compulsion to work, an inability to depend upon men to provide, and the public exploitation of women's bodies. Regardless of what form expression takes, the main point is that women's narratives are more comfortably distributed in private, face-to-face conversations in ways that keep information within the circles for which it is intended, narrowly defined. Public trust in the privacy of the Internet is not present and questions about governmental monitoring or a lack of anonymity linger in any user's mind. While this remains the norm, as Mona's narrative suggests, the Internet could change the shape and character of women's voices in the near future.

Mona

Mona is a college-aged Kuwaiti female who is at present studying in the United States. I conducted an interview with her online in October 2001. During our interview, I asked her how the Internet was affecting her life as
a Kuwaiti woman. She responded as follows:

> I usually use the Internet to do some research
> for school or else to chat with friends from back
> home. I'm not really into talking with a bunch of
> strangers. I don't really believe that gender makes
> a real difference. It really depends on what each
> person uses the Internet for.

When asked if the Internet had any special significance
for Kuwaiti women, she responded:

> I'm not sure if I see a major significance for
> Kuwaiti women. One thing that I see changing is
> that women try and do research on women's suf-
> frage, which is a major issue in Kuwait at the mo-
> ment. So women try and find a way to try and con-
> vince the government to let the women vote.

Mona's narrative reveals a tension that has been present in each of the interviews, a tension between tradition
and innovation. In this case, Mona states that she does
not like to "talk to a bunch of strangers," revealing a deep
cultural affinity for face-to-face conversation with people
one knows, loves, and trusts, such as family members
and friends. Reinforcing this perspective, Mona explains
that, for her, the Internet is a tool for keeping in touch
with friends from back home. Thus as a communication
device, the Internet reinforces existing bonds but does not
provide Mona with new contacts outside of her social cir-

cle. On the contrary, when asked if the Internet holds any special significance for Kuwaiti women, she observes that the Internet can be used as a research tool for female activists who are looking for data to support their arguments regarding the need for women's suffrage in Kuwait. Such activities represent a significant contribution to women's efforts to redefine traditional role models for women, in favor of a female civic culture. Mona's narrative suggests that some women in Kuwait are finding the Internet useful as a tool with which to enhance their role in public life. Her observations are characteristic of a different political environment growing in Kuwait, perhaps due in part to women's enhanced access to information and growing public voice.

Another small sign that women's attitudes toward the Internet may be changing is provided by student opposition to a government decision to impose gender segregation on campuses at Kuwait University. In 1996, Parliament passed a gender segregation law for Kuwait University, which immediately resulted in the gender segregation of cafeterias on the Idaliyya and Kaifan campuses and separate reading sections for women and men in university libraries. Islamic conservatives in Parliament, encouraged by these initial successes, continued their push for totally separate campuses and classrooms for women and men within five years (2001). The full implementation of the gender law ended a tradition of integrated education at the university stemming back to the 1970s. Some argue that separate campuses and classrooms for men and women will promote the education of Islamically conservative women. Without separate campuses and classrooms, Islamist families are unlikely to send their daughters to university. Critics of the law argue that the costs involved with building separate campuses and additional classrooms are astronomical and a waste of precious resources. Moreover, they argue that important life lessons are gained from coeducation. Conservatives argue that the majority of students at Kuwait University under coeducational circumstances attend the university not to study and learn but rather to interact with members of the opposite sex. Conservatives argue that segregation will improve the learning environment for all and thus is worth the investment.

Having gender segregation imposed on university students will make cyberspace an even more attractive meeting place for cross-gender interactions. Moreover, such rulings have encouraged students to use the internet to protest this law and its implications. In July 2000, a sit-in to protest the end of coeducation was launched. One of the leaders of the protest, a young woman, noted that the "non-aligned" student opposition movement intended to use the Internet to alert the world media and human rights organizations about its protest and about the ways in which the gender law violated women's rights in Kuwait. This new trend illustrates how the Internet can evolve into a tool for gender activism in Kuwait. It does not mean, however, that such activism will necessarily be effective at stopping or even slowing the trend toward Islamic conservatism in Kuwait.

Lessons from the Kuwaiti Case: Culture, New Technology, and the Persistence of Local Values

In Kuwait, local cultural constraints make female Internet use a limited force for social change, but recent data suggest that this may be changing. Most women with whom I spoke considered the Internet a tool for global feminist practices, divorced from, and unable to help with, the local struggles of Kuwaiti women.[12] Cultural hegemonies in Kuwait that define women's place, women's voice, and women's activism limit an open and organized gender struggle. The need to publicly link a text to a voice means that social pressures toward conformity for the good of the "family" (broadly defined) acted more powerfully to constrain women's voices than to liberate them. In Kuwait, a woman's reputation, and the ways that local information can be distributed to harm it, creates an institutionalized pattern for women's activism and voices. In spite of this norm, there is emerging evidence that attitudes toward the Internet may be changing, with more women using the Internet to struggle for women's right to vote and to oppose segregated university education.

Just because the Internet in Kuwait in 1996 when research for this book began, was not supporting open, organized, and sustained feminist resistance in a mass way does not mean that women were not an important part of Internet culture in Kuwait, or that their cyber activities were not having any social impact. As Carolyn Marvin observes, "Electric and other media precipitate new kinds of social encounters long before their incarnation in fixed institutional form."[13] This statement could easily be applied to the emergent Kuwaiti Internet culture and women's place within it. On the one hand, in 1996 we were not yet seeing the emergence of new "fixed institutional" relationships for women within Kuwaiti society (such as the vote), in spite of Kuwaiti women's active participation in cyberspace. By 2005, however, Kuwaiti women have learned to use their new communications possibilities "to put themselves on [a more] equal footing" with men.[14] In fact, as this book goes to press in May 2005, Kuwaiti women have succeeded in achieving full political rights. In part, their campaign was enabled through the effective use of (new) media technologies, including the Internet and text messaging.

As explored in this chapter, the Internet is supporting a whole range of "new social encounters" in cyberspace.

The Internet in Kuwait is enabling women to go to new places (such as chatrooms, where they can converse, unescorted, with members of the opposite sex) and to speak without having their gender influence the response of the conversant. It is enabling young couples to meet and choose mates independent of the family patriarchs. The Internet is opening up new employment opportunities to women such as Internet cafe ownership or home page design. It is enabling women to construct better informed arguments for women's right to vote, which by 2005 ultimately paid off. The Internet is providing a public forum for drawing world attention to student opposition to gender segregation at the university and the problems of growing Islamic conservatism in Kuwait. These incremental changes paved the way for significant social gains for women in Kuwait in 2005. For many women however, cyber practices continue to be shaped by the givens of Kuwaiti culture. Once the dialectic between old culture and new technologies takes root in local landscapes, however, we are likely to see those changes take on an increasingly fixed, institutional form.

Conclusion

This chapter shows how activism is shaped by local institutional and cultural imperatives, factors that discourage the majority of women in Kuwait from openly testing the chains that male hegemonies provide them. This case illustrates that just because capabilities to "know" and to "speak" are provided by the onslaught of new communications tools does not mean that such tools will be used freely without contextual constraints. Rather, complicated contexts of cultural, political, and social institutions seem to weave themselves around women's use of information in conservative Islamic societies. It is important to understand the lives and voices of women throughout the world, lest we make the mistake of thinking that "access" is the primary issue in building global solidarity with feminist consciousness and activism. Only by understanding the constraints upon women's activism can we know where to find, how to interpret, and how to encourage women's voices in the new electronic frontier.

The Kuwaiti case suggests that the Internet could enable increased gender equity for women by providing them with new professional opportunities sensitive to the cultural constraints of Islam and improved relationships between men and women by making communication across gender lines freer and less hierarchical. The Internet also gives women more frequent opportunities to interact safely with potential suitors. These "liberations" however occur against a background of continued conservatism, both politically and culturally, which limits the overt "revolutionary" impacts of the Internet on women's lives. The value structures of Kuwaiti society, although they limit overt feminist activism, provide the cement for holding society together. The uncertainties caused by the Iraqi occupation, and the social turmoil which was and is its legacy, keep many feminist activists more focused on processes of social healing rather than on their own "self-centered" advancement. The increase in drug abuse, divorce, violent crimes, unemployment, moral crimes, and the unhealed wounds of the Iraqi occupation, such as the continued complications of post-traumatic stress syndrome, help divide women's solidarity along social activist lines. The luxuries of life that many Kuwaiti women enjoy further divide women among lines of haves and have-nots, and also divide the haves along lines of beauty and appropriate social presentation. Differences in degrees of religious observance, as well as sectarian differences, also divide the power of women's voice. Thus one comes to understand women and the Internet in Kuwait against a complicated local background of cultural, social, and political givens that shape and limit activism, presenting a chorus of women's voices, each woman singing for herself and her community, although the tune may differ. The fact that in 2005 women have now gained institutionalized political rights may open up new avenues for women's leadership opportunities. In the meantime the Inernet and media communications give women new networking possibilities and ways for them to express themselves publicly.

CHAPTER 5

———

The Internet and Youth
Subculture in Kuwait

It has been written that "those who seek to understand what is happening in the Middle East today and to speculate about the area's future would do well to look carefully at youth, for they are the next generation in the process of becoming adults."[1] With this perspective in mind, several scholars in Kuwait have examined the relationship between youths and the Internet. In 1997, 1998, and 2001, three independent surveys conducted at Kuwait University found that nearly 75 percent of students surveyed were active Internet users.[2] In Kuwait, 57 percent of the population is under age twenty-five. Young people in Kuwait constitute both the highest concentration of Internet users, estimated to be approximately 63 percent of all Internet users, and the largest sector of Kuwaiti society (57% of the population was under age twenty-five as of 2001).[3] Moreover, young people's Internet practices are likely to stimulate the most significant changes in Kuwaiti society over time, for reasons explored later in this chapter.

Given Kuwaiti youth's sheer weight as a social force, their innovative communication strategies, and Internet savvy, as well as the fact that youth subcultures contain the seeds of future social norms, this chapter scrutinizes several young Kuwaiti descriptions of the importance and implication of the Internet in their lives. The most magnetic quality about the Internet which draws Kuwaiti youths to the technology, according to these interviews, is the way in which it enables the transgression of gender lines that are otherwise strictly enforced in Kuwaiti society. This capability is especially important for the majority of Kuwaiti youths, who attend or attended gender-segregated government schools.

In an attempt to examine those places and occasions where the Internet's cultural, social, and political implications emerge as public phenomena, this chapter suggests that Internet use by youths is creating new forms of communication across gender lines, interrupting traditional social rituals, and giving young people new autonomy in how they run their lives. Although these capabilities remain tempered by preexisting value systems, we are seeing important signs of experimentation that cannot help but stimulate processes of change over time as young people redefine norms and values for future generations.

This chapter is divided into two parts. The first reviews several unpublished studies of the development and impact of the Internet among Kuwait University students and balances these analyses against the author's own field research in Kuwait during the period 1996–1998. The second examines these findings in light of seven in-depth interviews conducted with Kuwaiti students who are active Internet users. Interviews were conducted online and occurred between 2001 and 2004.

Survey Research on Kuwaiti Youths and the Internet, 1996–2001

In March 1998, the Kuwait Foundation for the Advancement of Science and the Kuwait Institute for Scien-

tific Research co-hosted Kuwait's Conference on Information Super-Highway. This was the first professional conference held in the Gulf to consider the development and
impact of the Internet in Kuwait, and the Islamic world in
general. At the conference, most of the discussions of social impacts of the Internet in Kuwait focused, not surprisingly, on the youth subculture. Drs. Mazeedi and Ismail presented a paper which independently came to the
same conclusions as my own—Kuwaiti youths seem to be
the most deeply affected by the transformations in communicative practices enabled by the Internet. Mazeedi
and Ismail focused on the ways in which the Internet was
detrimental to face-to-face social ties between peers and
among family members. They argued that customarily,
young (and old) Kuwaiti men had gathered in the late afternoon and evenings to drink tea and eat sweets together. Likewise, women had done the same. With the introduction of the Internet, young people maintain that
they find it more enjoyable to surf the Net in the evenings
instead of participating in traditional social rituals. A regional survey of Internet practices in the Middle East
found that 55 percent of Internet use takes place between
the hours of 4 P.M. and 12:00 A.M., the hours when tea
and home visits, or visits to the *diwaniyya* (male social
clubs), are most likely to occur.[4] These survey results add
weight to the concerns of Mazeedi and Ismail regarding
the social effects of the Internet on Kuwaiti youths.

Mazeedi and Ismail also found that young people are
unlikely to use the Internet along with other family members (just over 10% did), which means that family ties are
potentially jeopardized by Internet use. Instead, youths
share their ideas and positive energies with people in cyberspace. While such interaction fosters a positive sense
of being one with the world (at least according to many
students I interviewed), it also opens up Kuwaiti youths
to contexts in which new thinking, perhaps contrary to

their upbringing, can grow unchecked by traditional authority figures. Only 7.8 percent of the students surveyed
by Mazeedi and Ismail were taught to use the Internet by
a family member. Thus the authors conclude that "families don't set the rules or standards on how to use the Internet ethically and academically."[5] Moreover, since students often use the Internet to meet with the opposite sex
(more than 30% admitted to this as a regular practice), Islamic sanctions against interactions with the opposite sex
outside of relatives and marriage are transgressed. In
some cases, cyber dating challenges the role of the family
as matchmaker.

My fieldwork corroborated the findings of Mazeedi and
Ismail, revealing that the Internet has become an important part of youth subculture, especially as a tool for
leisure and for communicating with the opposite sex.
While Mazeedi and Ismail had conducted a formal survey
to reach their conclusions, I based my analysis on a
smaller pool of in-depth interviews with students, computer lab administrators, and Internet cafe managers and
owners, as well as on participant observation. Irrespective
of differences in method, our analyses generated similar
data. My interviews and observations show that most students use the Internet for chatting. Youths often use the
Internet to communicate with members of the opposite
sex, while sitting in the same Internet cafe or computer
lab. At other times they contact acquaintances with
whom they have shared their nicknames (when meeting
in chat rooms) or e-mail addresses.

While many of the students I interviewed join international channels, they tend to communicate with people
with whom they share a common bond—such as being
Muslim or Kuwaiti or having a background in engineering, for example. Many of the women stated that they enjoyed talking with members of the opposite sex because
they did not generally have firsthand knowledge of how

men think. They explained that they valued gaining a male perspective on many of the problems they faced in daily life, such as fights with friends, tensions with parents, or concerns about the ideal spouse. To them, men were a mystery, outside of their interactions with family members, and the opportunity to interact with strangers of the opposite sex in a safe and an anonymous way was magnetic for them.

Throughout my fieldwork I met people who had fallen in love, or attempted to, through the Internet. One person I interviewed explained that her brother and sister-in-law had fallen in love via the computer. They met in a chat room. Over time, they developed a regular cyber relationship. One day, several months into the relationship, they decided to meet in person. When they went to pay for their Internet subscriptions at the Ministry of Communication building, they decided to wait for each other near the entrance. It was love at first sight, and after a period of courtship, they ultimately decided to marry. There were some problems, however, because he was Shi'i and she was Sunni, and her parents refused to bless the engagement. Ultimately, love won out, and their wedding cake was shaped like a computer, a symbol of the amazing tool that brought them together, enabling the transgression of sectarian lines dividing Kuwaitis and interrupting the ritual of arranged or semi-arranged marriages.

The Internet cafe owners and managers I interviewed observed regularly that computers and Internet connections were used mostly for chatting, and often with the opposite sex. The same was true for Kuwait University computer lab administrators, who often expressed disgust that students were not using the Internet as it was intended, to help with research and academic affairs. On several occasions, I was invited to join female acquaintances as they journeyed to meet potential suitors they

had met online. The meeting place usually was some-
where public, and neutral, like an Internet cafe. We would
meet outside, and then enter the cafe as a group. We
would sit around a table and drink coffee and get to know
one another. In a conservative Muslim country, there is
safety in numbers. As a foreigner, my presence also acted
as a buffer (perhaps the male at the table was with me).
The women who pursued these visits explained that such
liaisons were common.

Mazeedi and Ismail reached the conclusion that these
transgressions of gender lines and parental power represented immoral behaviors. My study argued, on the contrary, that cyber-relations could in fact help young men
and women in Kuwait understand each other in a way
that might improve communication and understanding
between the sexes in marriage and the family. Moreover,
the Internet might give young people more sovereignty
over the choice of a spouse. Often family members choose
a spouse for their child based upon standards related to
what might improve or protect the social status of the
family as a whole. Thus sometimes the question of love is
not a factor, and familiarity is not encouraged. Love, it is
said, will grow out of a good marriage. Ironically, or perhaps not, many young people I interviewed felt that parents were the wisest and best agents for determining the
suitability of a marital partner. They argued that unlike
matches made through the Internet, a family's choice was
determined by rational calculations rather than fickle
emotions or incomplete information.

Again revealing the conservative nature of Kuwaiti society, Mazeedi's and Ismail's survey revealed that 73.4
percent of students who use the Internet felt that it was
being used in socially abusive and ethically unreliable
ways. Similarly, 61.1 percent of those surveyed felt that
"the morals and behavior of the students have been af-

fected negatively by the Internet."[6] In 2001, a study by Professor Hassan Abbas at Kuwait University argued that students continue to misuse the Internet. Many of them "sneak into Internet cafes to freely browse through sights linked with immoral activities without check."[7] The guilt students feel over using the Internet to violate Islamic moral standards and the risks they take in transgressing social norms have not stopped cyber dating or visits to pornography sites from occurring.

While some Kuwaiti students are critical of the ways in which the Internet enables them to violate the norms they are raised to embrace, others are taking full advantage of the Internet as a vehicle for challenging Kuwaiti society's increasingly conservative view of proper public gender relations. Because the Internet supports practices that interrupt traditional chains of authority at work in their lives, students increasingly find cyberspace an attractive place in which to experiment with unfamiliar or endangered forms of social interaction. To a degree, students' cyber relations reveal that the Internet supports "decentralization, individual empowerment, resilience, and self-sufficiency," practices that coincide with the design principles of the technology.[8] The fact that many Kuwaiti youths remain critical of such practices illustrates how Muslim values help filter and buffer the meanings and implications of such experiences. Local cultural and social frameworks both shape what is revolutionary about the use of a new tool (transgender communicating is not a violation of any norm in the United States, for example) and influence the pace of change. The Kuwaiti case suggests, however, that even if Islamic sanctions on behavior exist, some interruption in value structure is likely when new patterns of communication are possible and experimentation persists, even becoming open and organized in temporary fits of passion.

The Internet and Its Impact—Testimony of the Participants

Scholarship on the role of the Internet in Kuwaiti youths' lives continues to stress that Kuwaiti youths are attracted to the Internet because of the ways in which the technology liberates the user from the conservative social codes of the Gulf. Survey data, however, suggest that identity continues to shape use, in that many of those surveyed feel discouraged to participate in such Internet practices as they threaten the norms and morals of Kuwaiti society. These same contradictions and tensions exist within the narratives analyzed later in this regard. Through these narratives, we see reasons for both processes of innovation and change, as well as the persistence of local cultural values and attitudes.

Fahad

Fahad is a Kuwaiti in his early thirties. He is completing a Ph.D. at the University of Leeds in the School of Computing. I met Fahad at an international conference in Cairo. We share research interests in the development and impact of the Internet in Kuwait. This interview was conducted via e-mail on October 21, 2001. His testimony offers a perspective on the Janus-faced nature of Kuwaiti attitudes toward the Internet.

The Internet is becoming *increasingly*, day by day, more dangerous and fascinating in our life. Do you believe me if I tell you my son, Muhammad, comparing to his age [eight], is an expert in the Internet? He can use most if not all the search engines; he has his own Web site on the WWW. He does 90 percent of his school homework by the Internet, and have five or probably six different e-mail addresses, and so on. At University of Leeds, there are around thirty Kuwaiti students; most of them (95%) are very young [ranging] from eighteen to twenty-two years old. They are crazy users of Internet. They use the Internet for different purposes—e-mail, chatting, sending messages, browsing the WWW, but mostly remains chatting, with at least three to four hours daily (the minimum).

This interview suggests the importance of the Internet in young Kuwaitis' lives. Fahad claims that Kuwaiti students at the University of Leeds spend at least three to four hours a day online, and that a majority of their online time is spent chatting. He argues that this is both "dangerous and fascinating," illustrating a bifurcated view toward technology and transformation. The language used to describe his son, Muhammad's, cyberpractices suggests fascination at the adeptness of an eight-year-old boy who is creating content for the Web as well as using it to enhance his intellectual life. More ambiguous are Fahad's attitudes toward Kuwaiti college students, whom he sees spending considerable amounts of time chatting rather than using the Web for serious research. In Kuwait, many computer lab administrators, as well as students I interviewed, described the addictive effects of the Internet on young people's lives. Those interviewed observed that performance in school was affected negatively by time spent online, chatting rather than studying. The effects of the Internet on young Muhammad's life, however, suggest that the Internet does not automatically detract from the educational process. In fact, at this time, the Ministry of Education in Kuwait is working to finalize the implementation of a project "on e-interaction," which will integrate computer training in the latest technologies into government schools.[9] The e-interaction program is part of the Kuwaiti government's effort to upgrade the

quality of public education. It could be that getting beyond the fascination with chatting will come from training
to use computers for more cerebral activities, as is the
case with Muhammad. At present in Kuwait, much training in the use of computers comes from friends teaching
friends how to use the technology. Often the first thing to
be taught is how to use the IRC (internet relay chat). I
was asked numerous times when in the computer labs at
Kuwait University if I could help a struggling student (a
"newby") to get logged on and to participate in a chat ses-
sion, but I was never asked to help anyone with a research problem.

Hassan

Hassan is a nineteen-year-old Kuwaiti who is finishing a B.A. degree at a university in Texas. He is the student of a colleague of mine, and he volunteered to participate in an online interview process designed to learn
more about the impact of the Internet on young people's
lives in Kuwait. Unlike Fahad, I have never met Hassan in
person. While the online interviewing process gives those
being interviewed a degree of anonymity and sense of
freedom to reveal their thoughts without self-censoring,
for the interviewer, there is no human connection with
the subject. Thus it is difficult to contextualize the narrative in a broader framework of familiarity, such as how
their voice sounds, their capabilities in spoken English or
Arabic, how they dress, what they look like, and so on.
One can deduce from Hassan's status as a Kuwaiti student abroad, as well as someone who attended a private
school in Kuwait, the American school at that, that his
parents are more than likely liberal in their politics and
relatively wealthy, and that Hassan is an above-average
student, as he was admitted to a well-respected U.S. institution of higher education. His testimony helps explain

why young people in Kuwait are interested in the Internet and provides some preliminary thoughts about the potential impact of the technology, both politically and morally. His testimony is unique in that his parents also are Internet users, again suggesting a fairly cosmopolitan and upper-class background. Most eighteen-to-20-year-old Kuwaitis that I interviewed had parents who did not use the Internet.

According to Hassan:

> I think the Internet provides an "escape" for youth in Kuwait. There isn't much to do in Kuwait other than going out to eat or shop and on weekends the beach house. The Internet provides uncensored news; one could argue it gives a different view of news around the world. It is for the most part up to date and current. I use it for e-mail and to check sports scores and the news. I also shop on the Net. For the most part, I use it to chat online with friends. For me it's never been hard to speak across gender lines. I went to an American school from K–12, and I guess I just got used to seeing and talking to girls every day. But I've realized that people who go to government schools (segregated schools) find it very hard to talk and socialize with girls. They just haven't had the everyday interaction with girls that my friends and I had. Politically, I think the Internet will help women to get the vote. It also provides a place where free speech can be expressed, since there's no limit to the Internet, and "policing" it is hard. Morally, the Internet does have some wrongs, pornography, which has been banned by a firewall in Kuwait, posed some problems in Kuwait. As for cyber dating and the like, I think the Internet is a "way out" instead of using the phone or trying to meet somewhere.

> For me personally, the Internet has made life eas-
> ier. In school, research meant surfing the Web
> looking for information I needed. Socially it has al-
> lowed me to keep in touch with friends from high
> school. Also, my parents use the Internet, so it has
> helped me to keep in contact with them.

Hassan's attitudes toward the Internet are cosmopolitan, like his background. For him, gender issues have not
been an issue, so he is less drawn to the technology as a
means for transgressing social codes. Instead he uses the
Internet as the average Western college student—for
shopping, chatting, research, news, and information.
There is nothing culturally specific about his testimony,
except what he is observing about others in his society.
Kuwaiti women, he notes, may use the Internet to advance their struggle for suffrage. Moreover, less well off
Kuwaitis who have not been afforded private prep-school
educations and instead have gone to government schools
that are gender segregated may be attracted to the technology in order to ease conversations with the opposite
sex. Hassan's testimony adds the important layer of class
background, which clearly influences how one views and
uses the Internet. At one point in the interview, Hassan
observes, "The Internet was 'the place to be' socially when
it was first accessible in Kuwait. Now, since almost everyone has it, people haven't paid much attention to it. It
was like a fad, it came and left."

The introduction of other communications technologies such as television, the cell phone, and the personal
computer has mimicked this "hype" then "massification"
then "normalization" model. It does not mean that the Internet "has come and left," to quote Hassan, as the technology has more users today than it did several years ago.
In fact, use of the Internet in Kuwait has grown an average of 14 percent a year for the past four years.[10] But the

hype associated with having an e-mail account or chatting seems to be waning for upper-class youth, who were
the first to have access and, subsequently, the first to
have the technology become a normal part of their everyday life. Hassan does not say that he does not use the Internet now, but rather that the mystique behind the technology that drew privileged Kuwaiti youths to the "new
toy" is waning. A similar observation was made by a
group of Egyptian students at American University in
Cairo. The students argued that the hype regarding the
Internet was a fad, even though at the same time the Internet had become a normal part of their day-to-day lives.
Their professor asked, "I thought you guys spent hours
on- line chatting?" They replied, "No, sir, that was last
year, chatting is boring."[11]

Hassan's testimony yields another important set of
tensions between the faddishness of new technologies
and the normalization and integration of new communication tools into everyday life. The transition from fad to
norm is linked to expectations of impact as well. Sometimes what turns out to be just a fad nevertheless initially
enables us to imagine sweeping change and a redefinition
of the world as we know it. When Al Gore was first promoting the Internet as a global communication and business tool, he promised that it would rapidly transform the
world. But such expectations are generally short lived,
and impacts from fads tend to be shallow and temporary
as well. Normal parts of everyday life instead conjure images of change that are incremental, evolutionary and
silent; however, it is the normalization of aspects of everyday life that provides the power for sweeping, mass-
based, and lasting changes that run through all sectors of
society, like capillaries in the body. What Hassan describes is a world in which the Internet is no longer hyped
but instead a normal part of daily life, likely to create
deep, long-lasting changes throughout Kuwaiti society.

Buthayna

Buthayna is a friend of Hassan's. She also is a Kuwaiti college student completing a B.A. degree in the United States. She is twenty years old. Part of my interview with her follows:

Q. Why do you think the Internet is so popular among Kuwaiti youth?

A. Well, I have been told that you have lived for a while in Kuwait, so I would gather you are familiar with the way in which the Kuwaiti society is built. There is a somewhat double standard, and there are many gray areas in terms of the two sexes mingling with each other. Therefore I think the most common place for both sexes to mix with each other is through the Internet. Girls especially cannot form relationships with boys, even as friends in many families in Kuwait, so the Internet is a "safe" place I guess for them to do so. And the fact that the two sides don't know each other, they feel safer to voice their concerns, ideas, without having their reputations ruined or without it affecting their social life.

Already in this interview, we note an interesting difference between the testimony of Hassan and Buthayna. Even though they come from similar class backgrounds and have similar educational patterns (both having attended private school in Kuwait), Buthayna, as a female, reaffirms the general pattern of young people's Internet use in Kuwait. She is drawn to it because it provides a neutral ground on which females can interact with males without fear of social consequences. A woman's reputation is something to be carefully guarded, and interacting too freely and openly with the opposite sex is a sure way to blemish one's social standing as a respectable woman. Men are not subjected to the same rules. If they talk to other women, it is the woman who is at risk, not the man. Thus the Internet for Kuwaiti women is a place where they can overcome this "double standard."

The interview continues:

Q. Do you think the Internet has any special significance for Kuwaiti women? How do you see the Internet affecting your life as a Kuwaiti woman? Does gender make a difference?

A. The Internet indeed is different for a woman than it is for a man, in many ways. As I have said earlier, due to the society that we live in, women are still bound by so many more rules than men are, even if people in Kuwait are not willing to admit it. Therefore, the Internet makes it easier [for] a woman to experience much of what she might not be able to experience in real life, even though this may just be virtual. In terms of research, it is also different, for there are many subjects in our society that are considered taboo, may they be sexual or not, so the Internet makes it easier to delve into many worlds, sometimes answering questions that cannot be asked, or just opening new horizons.

Buthayna reinforces the differences between female and male Internet use in Kuwait with the aforementioned observation. Moreover, she suggests ways in which Internet use can introduce subtle innovations in youth subcultural practices, making it easier for women to experience that which they might not be able to access otherwise, including the male psyche and taboo information.

Sabiha

Sabiha is nineteen. She is Kuwaiti and is finishing a

B.A. degree at a university in the United States. She also is a friend of Hassan's. Her testimony is similar in character to Buthayna's. She once again stresses that the major impetus behind Internet use among young Kuwaitis is the desire to communicate with members of the opposite sex. Illustrating the conservative nature of Kuwaiti society, Sabiha finds that there is a difference between chatting online (which is relatively harmless) versus having a relationship with someone online, which she observes "is impossible" as well as inadvisable, for the reasons that follow:

Q. Why do you think the Internet is so popular among Kuwaiti youth?

A. The main reason [the] Internet is so popular with the Kuwaiti youth is because it's the most effective way for boys and girls to communicate with each other. Mostly they use the Internet to chat with people from the opposite sex because, to them, it easier to communicate with a name and not a face. Very rarely, if ever, do they use the Internet to do any research.

Q. Do you think Internet use has a positive or negative effect on Kuwaiti youth? If so, in what way?

A. In some ways, there is a positive effect on the youth, because they are more able to communicate with guys, and it's a way for them to know that guys aren't so bad. The bad thing is that some girls try to have relationships with someone online, and that isn't possible. Many guys think this is possible and wind up having something like three or four if not more girlfriends online. Then there are the girls who try and do the same thing. Of course, this causes the problem that girls wind up not wanting to trust guys, and vice versa. So this is a major problem.

Cyber dating, although it is common, is viewed as "a major problem" by Sabiha. Because there are no face-to-face responsibilities between the parties and no firm commitment to monogamy, online relationships are said to be breaking down trust between the sexes. Chatting, on the other hand, is viewed by Sabiha as a positive way for girls to understand that guys "aren't so bad." This perspective is interesting, as it implies a degree of female solidarity as well as a gendered separation common in Islamic societies. Women's attitudes toward men in the Islamic world often are conveyed in terms one might see applied to foreigners. Men are clearly an "out-group," with strange thoughts, desires, and appearances. Only within marriage will these mysteries and a sense of foreignness be breeched. Cyberspace also provides some data with which to solve the mysteries of the opposite sex.

Sa'ad

Sa'ad is a college student at the University of Leeds. He is Kuwaiti and is completing an undergraduate degree in information sciences. Sa'ad is a friend of Fahad's. Sa'ad chose to begin his narrative with a description of the Internet's place in Kuwaiti society. He observes:

As a start, I'll give you a brief status of Internet use in Kuwait.
1. [The] Internet is used in Kuwait mainly as an entertainment source (chatting).
2. Internet cafes are widespread through the country and they offer Internet service 24 hours, seven days a week for about $1.50/hour.[12]
3. The age group of people who use Internet

ranges between thirteen and fifty, but it's generally between fifteen and twenty-two.

When asked why the Internet is viewed as a useful technology to young people in Kuwait, Sa'ad chose to respond in terms of his own experiences.

> Personally speaking, I became interested in the Internet when I was fifteen or sixteen; I used to spend hours and hours on the PC day and night chatting with people on the Net. It was an easy, an effective way to find cyber friends [who] share the same thoughts and ideas. As [you] may know Kuwait is a conservative country and when coming to relations between genders it's a bit like a taboo or something forbidden before marriage. By using the Internet it made it easy to cross the prohibited barrier and to allow communication between the genders freely. When I was young I had a problem of expressing my thoughts and feelings because I was shy. On the Internet I could do that so easily without any embarrassment.

Once again, Sa'ad's observations confirm the importance of gender codes and the ways in which cyberspace enables their transgression as a motivating factor for young people's Internet use in the Gulf. Sa'ad also adds another variable of individual personality when he confesses that he is a relatively shy person, and that he found it enabling to be able to communicate with people with a few degrees of separation.

> Q. When Kuwaiti youths use the Internet, for what do they use it?
> A. In Kuwait we have a saying, which in English means "every thing forbidden is wanted." Youth in

Kuwait use it for chatting and to mix with the other gender freely without any of them knowing who is behind the screen. As for youths whom their parents are strict and never let them out of [the] house or touch the phone . . . then they would go on the Net and talk while the parents are fooled that their son or daughter is doing the homework on the Internet. One more thing that is forbidden in Kuwait is pornography. Youths have always been curious and been eager to know or learn something about sexual relationships. It was never discussed between parents and their children, and if it was brought about between youths themselves, it was discussed in a wrong way.

Sa'ad's narrative suggests the possibility of major social transformation in Kuwait as a result of young people's Internet use. Through the Internet, youths are regularly interacting with the opposite sex and obtaining information that in the past had been considered taboo or immoral. Sa'ad says of this trend:

I would agree to a certain extent that [the] Internet is good for both genders to communicate, because it would help in [the] future when both have to deal in university or work. However, it wouldn't be possible to know who the youths are talking to or what sort of mentality the [people they were] talking to [had]. Of course, [no one] would want his son or daughter to chat with a rapist or drug user.

One important cultural variable at work in the Islamic world is the desire to verify someone's moral right and the authority to offer an opinion or a judgment, as well as a desire to assess the soundness of the advice or opinion offered. The soundness of the opinion is often judged in part by the character of the speaker or writer. In cyber-space, "no one knows you're a dog," as the popular *New Yorker* magazine cartoon observes. For Kuwaitis, and other Muslims alike, one of the problems of cyberspace is that it interrupts traditional systems for awarding authority and authenticity to public discourse.

This concern, however, has not slowed the Muslim use of cyberspace. Some have even gone so far as to argue that cyberspace has created a form of Muslim renaissance with the creation of online *fatwa* services, question-and-answer sites with *imams* from all walks of life and geographic locations, and Muslim community services directories, with offerings as diverse as where to buy *halal*[13] products to dating services to *madrassa* services and discussion groups. Sa'ad's comment that the Internet is both good and bad because it promotes dialogue across gender lines as well as discussions of sensitive issues such as sex but at the same time does not allow one to know the reliability and character of the person with whom one converses, is typical of the pattern of discussions that I have had with young people throughout the Muslim world, including Kuwait, Syria, Egypt, Morocco, Tunisia, and the UAE.

When I asked Sa'ad how he expected the Internet to affect Kuwaiti society, explaining that, as illustrated in the first part of this chapter, some scholars felt that the Internet threatened the social fabric of Kuwait, he responded:

> I can remember when I was young I barely used to go out with friends. I was cut off from the outside world and I had my own world on the Net. This affected my social [life], and I've lost contact with my friends. Surely it would have an impact on the political life, because the young wouldn't be spending time on reading newspapers, watching TV, or going to *diwanniyat* to discuss what's happening

recently in the country or world.

When I also explained that other scholars were concerned that online dating could undermine the role of the parents as matchmakers for their children, Sa'ad responded:

> My old flat mate, he's Kuwaiti. He met a Lebanese girl on one of the chatting rooms, and he had a relationship with her. They exchanged photos and phone calls after six months they met in Lebanon, and he met her family and they got engaged. Now they're happily married for almost two years. Now in Kuwait it's not necessary that parents have to find the perfect match. Their children can go and look themselves, and the parents might approve of their decision or not. Some conservative families would not allow this and prefer their children marry whoever they think is suitable. I certainly don't want to get married to someone I don't know, and I prefer to choose my match. Moreover, I would want my parents to have an opinion in my match but not to take the final decision. Marriage is an extremely serious thing, and I wouldn't like to make a decision just while chatting on the Net. I wouldn't mind getting engaged to someone I chatted with, but it has to go through the old-fashioned way of talking on the phone and face-to-face conversation.

Sa'ad's observations resonate with the concerns of other Kuwaitis who prefer the traditional ways of courtship, even if the Internet provides the initial contact. Most Kuwaitis I interviewed saw the Internet not so much as replacing the role of the parents in the matchmaking enterprise but instead broadening the pool of potential suitors. In commenting on the way dating generally occurs in liberal families, Sa'ad observed:

> The usual way of dating in Kuwait starts with exchanging phone numbers between a boy and a girl who may meet in a restaurant or in malls. Then phone calls. So basically the two exchanged the number just because they liked the way each of them looks like. However, when a couple date on the Internet, they've chatted and they knew the personality of each other. I may consider cyber dating is much better when the persons who are dating are serious and think of having a relationship not just for fun . . . however, it's still a risk to believe everything said on the Internet and to trust someone on the Internet to go out with.

Sa'ad's ambivalence continues to express itself in these observations. He feels that just to judge a suitor by looks is not good, and this is the way that courtship generally precedes in Kuwait. On the Internet, cyber daters get to know the personalities of potential suitors first, and this is something Sa'ad values. The difficulty in verifying the relationship between what one says in cyberspace and one's character, appearance, and behavior in real time and place makes cyber relationships "risky" in Sa'ad's opinion. In any case, parents are still envisioned as playing a role in choosing or verifying the choice of a spouse. The ambivalence, the preference for tradition, and the celebration of cyber capabilities all illustrate a traditional culture's attempt to adapt new and foreign technologies to the practice of everyday life. Preexisting cultural frameworks act as filters to the kinds of behaviors and capabilities deemed appropriate or useful. The syntheses that result provide a uniquely Kuwaiti response to the digital age.

Alia

Alia is nineteen. She is Kuwaiti and is a college student, at present finishing a degree in Nutrition at West Virginia University. She is a friend of Sa'ad's. Alia's narrative continues to illustrate the ambiguous feelings that the Internet conjures up among Kuwaiti youth. She observes:

> In my opinion, the Internet issue is really difficult to talk about because of the touchy subjects it holds. As in many other countries, the Internet is misused in Kuwait a lot. In my opinion, chatting on the Internet can be very damaging in the Kuwaiti culture, because many people who chat on the Internet use it as a tool to meet people of the opposite sex.

Once again we see the expression of conservative Muslim values in terms of a view that the Internet is harmful to Kuwaiti culture as it enables the violation of strict gender segregation that is the norm for Kuwaiti social life. Alia explains that the threat of the Internet comes from the fact that most Kuwaiti youths are interested in the technology because it allows such transgressions. When asked why the Internet is popular among young people in Kuwait, Alia stated:

> I think it's because, like I mentioned it's a good way of talking to the opposite sex without getting "caught" by their parents. Most of them just join chat rooms. Hardly any of the young people I know use the Internet for academic purposes or just for the purpose of looking up information. The most used chat program is mIRC. In the past, mIRC wasn't as bad as it is now. People still

joined to talk to the opposite sex, but at least there were limits to what you can talk about in a public channel. Now, people talk freely about whatever they want, and some people talk about vulgar and unwanted topics. It's now known that any girl who joins mIRC is not a "decent" girl and is only joining because she's looking for a boyfriend. Of course, since I'm a girl myself, I completely resent this idea. But just the fact that people think this way made me quit mIRC. And, just to satisfy my curiosity, I joined once and tried to have a conversation with a girl on mIRC, but hardly *any* girl wants to talk to another girl. That really disappointed me, because it proves that they really are joining just to talk to guys. That's another reason why I decided to quit. Another thing is, the age group keeps getting younger and younger. Now even ten-year-olds join. There is no parental control over this, which I believe is wrong, because these parents are leaving their children in the hands of strangers on the Internet, and most of them are very corrupting people!

Alia's narrative provides an intimate view of the way in which a woman's life is compartmentalized into a behavior that is safe for one's reputation—or not. Alia says that she "resents" the double standard applied to females who are discouraged from using new technologies, in this case, the Internet, in order to protect their "decency" at the same time that such standards are not applied to men. In a demonstration of experimentation and defiance, she tries mIRC anyway, to find out for herself what all the trouble is about. She finds it hard to get other women to chat with her and concludes that mIRC is "corrupting" since it really is designed to transgress gender boundaries. She voices her concerns that parents are not careful enough in monitoring their children online and that the age of participants keeps getting younger and younger. She once again voices distrust for strangers, which fits a common cultural pattern, and also concludes that most of the conversations in the chat rooms at present are "vulgar" and thus inappropriate. She notes the devolution of the utility of the tool, that it "wasn't as bad as it is now." Her narrative illustrates an observation made by Sa'ad, that in Kuwait, "whatever is forbidden is desired all the more." The more taboo a topic, the more likely it is to show up in chat rooms. The more Kuwaiti society tries to separate the genders, the more likely the Internet will be used to transgress such boundaries.

When asked about the potential long-term effects of Internet use by young people in Kuwait, Alia observes:

> Most students join chat rooms after their day is over. It does consume their studying time. I admit to wasting a lot of time on the Internet when I could have been studying. But as for social issues, I don't think it really gets in the way of those unless the person is *really* addicted. About the marriage issue, well, I might have a different view on this than other people, because I consider myself a bit more open than the average Kuwaiti. I've been raised in an English school all my life, and I believe that played a part in giving me different views than Kuwaitis who have been raised in government schools. Also, my parents raised me to believe that I don't have to marry someone through an arranged marriage. If I do happen to want to marry someone, of course my parents have to approve of him, but they don't have to select him for me. In addition, I believe many of the people who chat on the Internet are not chatting to look for a marriage partner. They are just looking for something tem-

porary. Especially men, many of them believe that the girls who join the Internet are not decent women, so how could they marry them? It confuses me, because they tend to contradict themselves a lot!! This is why I like to keep myself away from these things.

Alia again expresses frustration over the double standard that characterizes expectations for young women as opposed to young men. She also exhibits the interesting contradictions that characterize the identities of upper-class Kuwaitis, who are raised with cosmopolitan values and exposed to new technologies, yet heavily influenced by the conservative social values of the general context in which they live. Alia describes herself as open and liberal, but she still expects her parents will have veto power in terms of the spouse she selects. She wants to keep her reputation as a "decent" woman, yet she also wants to explore new technologies such as the Internet. Just because many people use the technology does not mean that she will. But given the general pattern of Internet use in Kuwait, her reputation will not be immune if she is viewed as Internet active, nor will those she meets online take her seriously as a potential spouse, since only indecent women are seen as lurking online.

When asked whether or not she finds anything morally wrong with the Internet, Alia replies:

In many ways I do believe that it's wrong, because they are doing these things behind their parents' backs, and they know it's wrong but they do it anyway. They also cyber date "undercover" so nobody knows who they are, and that way their reputations won't get hurt by it. I believe this is extremely wrong and immoral. Plus, how could you fall in love with someone you have never seen or met personally? How do you know they're not complete liars? It's very easy to lie on the Internet, and many people choose to ignore that fact.

Again we see the preference for face-to-face conversation, as it is possible to lie in cyberspace, and there is no system of verification or authorization. We also see an expression of local cultural values, which says that parents know best and that anything done behind their backs is more than likely immoral and unacceptable. In spite of these judgments, Alia values the place of the Internet in her life. When asked how the technology has influenced her life personally, she observes:

> The Internet changed my life a lot. I'm sorry to say that it was the thing that taught me all the bad things that people shouldn't do. I am lucky that my parents taught me better than to believe everything I see or hear from people, because if it weren't for that, I would be just like all those other girls who join the Internet to meet guys. It also wasted a lot of my time when I first moved to the United States, because I got very homesick and used the Internet *too* much! I could have utilized that time to study instead of chat with strangers! Nowadays I only use the Internet to check my e-mail, do research for classes, [and] talk to my family and friends on MSN Messenger.

As was the case with Hassan, Alia describes a common pattern of Internet hype and experimentation, transforming culturally acceptable use and the normalization of the technology into a part of day-to-day life. Long-term use patterns mimic both the social design of the context and the user. For Alia, long-term use of the Internet will help with her studies and will keep her in touch with fam-

ily and friends and people she already knows, even if in
the beginning she experimented, at great risk to her reputation, with "bad things that people shouldn't do."

Ghada

Ghada is a Kuwaiti high school student. Our connection is a result of the Internet. She contacted me in April
2004 to express her thoughts about an article I wrote
about the Internet and the Kuwaiti youth subculture. She
had her own ideas on the subject and asked if I would be
interested in hearing her thoughts. I said that I would,
and in our second round of e-mail correspondence, she
shared the following narrative:

> I think Kuwaiti people are extremely sexually re-
> pressed. Since they were born, boys were made to
> play with boys and the same goes for girls. Fortu-
> nately, I attend the coed American School of
> Kuwait. Unlike some Kuwaiti girls, I interact natu-
> rally with the opposite sex. Many of my friends in
> public schools become flustered and extremely shy
> when they talk to men and guys, even if they were
> doing a mere transaction at the bank or whatever.
> Nowadays, many girls and guys have a relation-
> ship through the phone. . . . Dating in Kuwait is a
> cultural no-no. And as you know, Kuwait is really
> tiny, and talk spreads like wildfire, and Kuwaiti
> people are naturally gossipy; it's in their genes.
> Speaking of the Internet, my guy friend in high
> school told me that last week his seventeen-year-
> old friend (a female) had a "relationship" with a
> twenty-one-year-old guy through mIRC (chat
> room). He told her that he loved her . . . and basi-
> cally the twenty-one-year-old guy convinced her to
> have actual sex with him. Afterwards, he dumped

her and carried on with his life while she, a seven-teen-year-old high school girl, was left in the disaster that happened. If a Kuwaiti girl loses her virginity before marriage, she's basically doomed forever and will not be able to marry because she is considered a prostitute or worse.

Ghada's frankness and transparency add clarity to the other narratives contained in this chapter. For example, through Ghada's words we not only sense, but touch directly upon the concerns (even among students themselves) regarding cross-gender interactions in Kuwait. Her descriptions of the differences between those students who attend coeducational institutions and those who attend segregated government schools are stark in terms of cross-gender interactions. Moreover, Ghada's frank commentary on the importance of a woman's virginity before marriage illustrates why some Kuwaitis may be concerned about cyberdating. She directly illustrates the cause for concern when she tells the story of the young woman who was pursued, courted, and blemished via a relationship begun on mIRC. It is this fear of "virtual" relationships becoming "real" relationships that is the main cause for concern, especially in light of the conservative and potentially unforgiving nature of the tiny glass house that is Kuwait.

Conclusion

This chapter examined those places and occasions where the Internet's cultural, social, and political implications emerge as public phenomena in young Kuwaiti lives. Using Kuwaiti students as a barometer for the future, this chapter illustrated that Internet use by youths

is intensifying communication across gender lines, interrupting traditional social rituals, and giving young people
new autonomy in how they run their lives. These narratives support the findings of surveys that have been
conducted by analysts from 1996 to 2001. Some important findings should be highlighted. First, the Internet
means different things to young Kuwaiti women than it
does to their male counterparts. Women are not afforded
the same freedom to talk across gender lines in public because of the need to preserve their reputations. Thus the
Internet is seen as a useful tool, even among women from
liberal, cosmopolitan backgrounds, as it safely allows the
transgression of gender boundaries, that is, until the
boundaries between virtual and real relationships are
breeched as illustrated by Ghada's narrative. Second,
while many Kuwaitis are concerned about the potential
misuses of the technology and are vocal about exactly
what these misuses constitute, these concerns have not
stopped them or their peers from being Internet active.
Third, preexisting value systems help shape long-term
use. This means that even if experimentation occurs, in
the end many Kuwaitis adjust their Internet usage to be
compatible with their upbringing and the norms and values of their society. Still, we are encouraged by these narratives to wonder if important signs of experimentation
cannot help but stimulate processes of change over time
as young people redefine norms and values for future
generations. At present, however, the weight of contextual
variables, from gender to class to ideology to culture, continues to strongly influence Internet practices, even
among the young.

CHAPTER 6

The Internet and Islam in Kuwait

Introduction

On January 29, 1997, local Kuwaiti newspapers in both Arabic and English ran front-page stories reporting the results of a *fatwa* issued by a sheikh in al-Jahra, one of Kuwait's most conservative towns. The *fatwa* demanded that local citizens rid their households of satellite dishes, VCRs, televisions, and video cassettes, all of which were determined by the sheikh to be *haram* (forbidden). A color photograph of a young boy standing next to a large pile of charred communications technologies accompanied the article.[1] This image suggests that for some Kuwaitis, albeit a small minority, a no-compromise policy exists when Islamic values are confronted by Western media discourses and the technologies through which they flow. From this perspective, the best way to maintain one's Muslim identity is to lead a simpler, less technologically engaged existence.

The rise of the Taliban in Afghanistan drew worldwide attention to the rejection of a Western-transmitted global culture through a campaign to round up televisions and satellite dishes. The Western media used images of the

163

Taliban putting television on trial to bolster a common story line representing Muslims being caught in the tidal wave of Western-inspired technological progress, for which they are said to be underprepared. Bobby S. Sayyid captures the spirit of this story line when he observes that "Muslims":

> are often thought to be out of time: throwbacks to medieval civilizations who [sic] are caught in the grind and glow of "our" modern culture. It is sometimes said that Muslims belong to cultures and societies that are moribund and have no vitality—no life of their own. Like ghosts they remain with us, haunting the present.[2]

Yet even for the Taliban, the Internet is different.

Taliban television wars resonate with the Kuwaiti sheikh's *fatwa,* and together they create an interesting contrast with Islamically conservative attitudes toward the Internet. When I learned that the Taliban had its own Web site (at least it did in 1998), I became curious about this apparent schizophrenia. The Internet seemed to offer a much wider range of access to un-Islamic materials and also gave the individual much greater control over what to access. In order to probe this tension, I decided to try to engage the Taliban myself, using of course, its e-mail address posted on the Web site (*http://www.taliban.com*). I posed the following question, addressing it to Abu Mujahid, as instructed (*abumujahid@taliban.com*):

> Dear Abu Mujahid,
>   I am a professor who teaches Islamic subjects at the University of Washington. I am presently writing an article entitled "Islam and the Values of Cyberspace." Many aspects of Internet life seem incompatible with Islamic values. Islamic

movements, and Muslims in general, however, seem to be using the Internet to spread Islamic awareness. I was surprised to learn that there is a Taliban Web site, because of your movement's attitude towards Western communications technologies like television and satellite dishes. So, my question is, "Is the Internet compatible or incompatible with Islamic values? Do the people of Afghanistan have access to the Internet if they want it, or is it mainly a tool for *da'wa* abroad?"

Within 24 hours, an answer arrived:

Dear Wheeler,
    Actually, there is some difference in [sic] TV and Internet. We have not [sic] control on TV broadcasts. We are compel [sic] to see everything without filtration. If there is any un-Islamic program, and there are almost all un-Islamic programs, so there is only one way, to turn it off [a resounding similarity to the al-Jahra sheikh's *fatwa*]. At this time, we have not capability to broadcast any Islamic program, so we ban TV here. But on the Internet it is up to the user. He can only see what he want and also easily filter unwanted sites so Internet is not only allowed in Afghanistan, but also will be used in many fields and developments in the future.

Two years later, the whole world would learn just how different the Internet could be for a conservative, and an increasingly militant, Islamic movement. Al-Qa'eda, an ally of the Taliban, would use the Internet, as well as a host of other western communications technologies such as cell phones, jet planes, fax machines, and credit cards, to commit one of the biggest terrorist attacks in history. Nearly 3,000 people would lose their

lives in this demonstration that there is no incompatibility between religious commitment and digital media. The
two fit like hand and glove. One lesson from September
11 that should not be lost in the flurry of media images
is that Muslims are not medieval throwbacks, underprepared for the twenty-first century. Just because they reject certain values and artifacts of Western culture does
not mean that they are not prepared to use our other
tools to promote a well-defined, populous message, and
sometimes violently. Confirming this view, a U.S. coun-
terterrorism expert, Roger Cressey, observed, "We under-
estimated the amount of attention (al-Qa'eda) was pay-
ing to the Internet. . . . They spent more time mapping
our vulnerabilities in cyberspace than we previously
thought."[3]

Another indirect lesson of September 11 regarding
Islam and politics, and reinforced by this chapter, is the
fact that the majority of Muslims reject the use of violence
to express themselves politically and religiously. Many Islamists have a passionate dislike of Western values,
which they see as blindly capitalistic, radically individual-
istic, and morally lacking. In spite of these strong feelings, the majority of Islamists choose to express such rejection discursively, with diplomacy, a pen, a keyboard, a
demonstration, or a mosque loudspeaker, or symbolically,
through, for example, a grassroots boycott of Coca-
Cola™, McDonald's™, and Disney™ products.

The Kuwaiti case of Islamic uses of the Internet illustrates aspects of Islam's compatibility with the Internet, as well as the ways in which this union aims at local social reform, while refraining from overtly
revolutionary (i.e., attempting to overthrow the government) or militant activities (such as terrorist attacks).
The following pages examine these expressions in
Kuwaiti cyberspace, especially in the fields of finance,
medicine, and *da'wa* (spreading Islamic awareness and

conversion). Throughout this analysis, I attempt to illustrate that media representations of Muslims as being backward and/or violent grossly underestimate the power, skill, and capability of Islamist movements in the world today. Examples of the growing power of Islamist movements in Kuwait also are considered in light of the efforts to use the Internet to raise religious observance and awareness in the areas of banking and medicine and efforts at conversion directed at expatriate communities in Kuwait (roughly 55% of the population).

The Power and Influence of Islamist Movements in Kuwait

In a rare moment of transparency, and speaking from the heart, Sheikh Saud Nasir al-Sabah, former Kuwaiti Ambassador to the United States, former Minister of Information, and former Minister of Oil, offered during an interview with the *Sharq al-Awsat* newspaper the following description of communal life in his country: "The Islamists have hijacked Kuwaiti politics."[4] Sheikh Saud's assessment on the surface depicts the general liberal consensus regarding the present Kuwaiti political environment. Carried by wire services around the globe, Sheikh Saud "called on the country's silent majority to rise up and quell the growing Islamist influence" in Kuwait.[5]

In part, Sheikh Saud's message responds to the enhanced activism of Islamist elements in Kuwait. Thumbnail sketches of this activism include a parliamentary effort to align Kuwaiti legislation with the Shariah, to introduce bills in Parliament to segregate Kuwait University, to censor the Internet and to ban public concerts and fashion shows. Reinforcing this trend was the discovery that one of al-Qa'eda's chief organizers and spokesmen was a Kuwaiti (he has now been stripped of his citizenship), Sulaiman Abu Ghaith, a former teacher and mosque preacher, who remains one of Bin Laden's most important ideologues. Besides reflecting Kuwaiti politics, Sheikh Saud's words contain a subtext of his own personal confrontation with Islamist political forces. In 1998, a struggle with Islamists in Parliament eventually cost him his post as Minister of Information.

In contrast, political scientist Dr. Mary Ann Tétreault observes that "the fearsomeness of Kuwaiti Islamism today revolves around its surprising success at imposing gender segregation at postsecondary schools in Kuwait," and that "this victory is interpreted as a reflection of Islamism's hegemonic power."[6] When one looks beneath the surface of this highly public success, Tétreault notes:

> There are few other issues capable of mobilizing most or all of Kuwait's Islamist political forces on the same side. As the "easy" issues are used up, Islamist cohesion and Islamism's political base are likely to erode.[7]

Tétreault bases her assessment on the argument that "neo-fundamentalist attacks on civil liberties already have begun to edge into danger zones for Islamists leaders, few of whom seem to appreciate the degree to which modernity has penetrated Kuwaiti life."[8] To characterize the diversity of public frustration with some aspects of Islamist politics in Kuwait, Tétreault observes:

> Demands to prohibit mixed-gender public gatherings such as concerts and lectures stimulate backlash among students. Calls to ban books and films are resented by middle-class intellectuals. Proposals to outlaw satellite dishes might even mobilize housewives angry to lose access to their favorite foreign programs. . . . The fragmentation

characteristic of modernity disrupts Islamist plans for social control by diminishing the number of citizens who share their monadic worldview, even from within the Islamist movement itself. [9]

Based upon this image of fragmentation that results from more intricate schemes by Islamists to control Kuwaiti politics and social practice, Tétreault concludes, "Islamist successes themselves plant seeds that undermine future political victories."[10]In other words, Islamist politics in Kuwait are not a sustainable enterprise. By implication, modernism and liberalism will survive as the dominant political culture of Kuwait.

If Islamists have hijacked Kuwaiti politics, then it is just a temporary blip on the democratization radar screen. While there is something comforting in Tétreault's assessment for Western and liberal observers, the increasing sense of siege and shrinking public space for liberalism and the disorganization and sense of powerlessness, "underfinancedness," and lack of new vision that many liberals in Kuwait express add weight to Sheikh Saud's assessment. As this chapter argues, the power and influence of Islamists in parliamentary life, Islamist information strategies, and increasing evidence of Islamic observance in social practice in Kuwait all attest to the growing place of Islam in Kuwaiti public life.

Promoting Islamist Discourse

In addition to efforts to limit public discourse that is not "Islamic," Islamist movements in Kuwait pursue active campaigns to saturate public space with signs and symbols of a faith-centered existence. Islamists have campaigned to have more radio space dedicated to Quran recitations and other religious programming. Islamists have helped develop television shows, including the popular Saturday afternoon broadcast of an "ask the Imam"-style show hosted by Sheikh Khalid al-Mathkur, a member of the Amiri Diwan and head of the Committee for the Implementation of the Shariah.

Some Islamist activists own and operate their own Islamic bookstores, which actively distribute both local and international writings on proper Muslim conduct, as well as classical *fiqh* books, science and technology writings in Arabic, cookbooks, Qurans, children's literature, and educational materials. Also sold at these bookstores are Islamist symbols with which to decorate one's car: bumper stickers with the name of God and/or his prophet (in Shi'a bookstores, stickers with the name Imam Ali and his son Hussayn in beautiful calligraphy are common), small Qurans or icons to hang from one's rearview mirror, and so on. Quran recitations in audiocassette and CD form as well as some software are sold at these bookstores. The most popular-selling items are contemporary religious writings including books by Sheikh al-Qaradawi, collections of *fatawa* for Muslim women, and books on marriage and family issues, prayer, and Islam and science. Islamist bookstores are gathering places for intellectuals and activists as much as they are places for distributing information. Some Islamist bookstores are located near conservative women's clothing stores where Islamic dress, including the *a'bayya* and *hijab,* is sold.

Another place in which Islamist discourse is promoted is in the local press. Local Kuwaiti newspapers increasingly provide space for Islamists activists to express their views. In the *al-Watan* newspaper, a special weekly insert covering religious issues and politics is printed and distributed on Friday. Islamists also have their own newsmagazines and journals, hold meetings and conferences to discuss various social and political issues, and organize a variety of activities targeted at different audiences, including children's programs, programs for students, activities for the elderly, and programs for women. Mosques also are important centers for the distribution of Islamist discourse.

The most general and public effort to promote Islamist discourse in Kuwait is the phenomenon of Islamist road signs. Posted illegally without license from the state, various Islamic organizations in the late 1990s launched public information campaigns in which simple slogans such as "trust God" or "there is no god but God" and "Muhammad is the prophet of God" or "God is great" were placed along highways, in the gardens of roundabouts, and along neighborhood roads. Signs were especially posted in strategic, congested spots where drivers were likely to slow down or merge into traffic. Signs were also posted along the most highly traveled roads such as the ring roads or the major highway links between neighborhoods. A Ministry of Public Works employee during an interview explained that technically such signs were illegal, and that any road sign had to be licensed by the state before being displayed. He explained how the government had invested millions of dollars in finding a material for overpasses and bridges, to which no public bills would effectively stick, in an effort to manage public information. As a result, Islamist road signs often were rounded up and taken down. They came in waves. One night some organization would go on a public campaign to place signs in town along roadways. They would last for a week or two, and then someone would take them down. A few weeks later, new signs would appear in different locations. Most of the signs that were officially licensed by the government had a commercial purpose, such as advertising fast-food restaurants, auto dealerships, or computer companies. The only official government signs that incorporated a religious message included two in the heart of Kuwait City, one that stated that it was imperative that Muslims protect their families by wearing seat belts, and another that said that it was against Islam to litter.

Whether through road signs or newspaper columns, bookstores or mosques; clubs for women or children, or by bringing liberals to trial for their "blasphemous" words, Islamist organizations employ multilevel strategies to promote their control over public life. At the same time Islamists use their power in Parliament and the institutions of the court to police public space in the university and in society in general to limit types of politics and social conduct that are viewed as un-Islamic. In their efforts, Islamists present an organized and a powerful message to a public increasingly searching for answers to questions regarding government corruption, increased economic hardship, lack of career opportunities for young people, as well as increased public dismay with what is viewed as a decline in the social and moral character of Kuwaiti society.

Islam and Cyberspace in Kuwait

Early in the Internet's history, observers expected that because of the technology's design structure, the Internet would promote processes of liberalization, as an open flow of communication and opinion was enhanced. But as Mark Poster argues, "The only way to define the technological effects of the Internet is to build the Internet—to set in place a series of relations that constitute an electronic geography."[11] In the Kuwaiti case, the electronic geography of the Internet, as explored later, supports increasingly overt signs of Islamic conservatism. Given the fact that Islamists are highly organized (Shafiq al-Ghabra calls them the only unified, well-funded, and organized force in Kuwaiti politics and the Middle East as a whole), highly educated (many are engineers or scientists), highly motivated (may have a clear vision of how they want to change the world), and skilled in IT, it is not surprising that they are a defining force in Kuwaiti cyberspace. If we lay aside presumptions that Internet technologies de facto promote liberal democracy, and the notion that Muslims are technologically relatively unskilled (a common story line in Western media), then we will more accurately see how the forces of globalization, blended with local cultural identities, utilities, and imaginations, produce a synthesis of modernity and tradition. This synthesis of technology and conservative religious identity is poised to fill the power vacuum left by the bankruptcy of liberalism, nationalist-based authoritarianism, and failed economic strategies in the region.

A Web Site for Islamic Banking

The Kuwait Finance House is a financial management company that builds assets locally and globally and provides traditional banking services, governed by strict Islamic principles—including observance of the ban on usury. (The Web address for the Kuwait Finance House is *http://www.kfh.com*). The Kuwait Finance House is not the largest bank in Kuwait. It is dwarfed in size by the National Bank of Kuwait. But it is the only bank in Kuwait that is governed by Islamic financial principles. The Kuwait Finance House (KFH hereafter) is located in the heart of the Kuwaiti financial district, downtown, near the Ministry of Information. The building that houses KFH, a tall mirrored, glass skyscraper, symbolizes the merging of Islamic principals and modern technologies. The KFH was one of the first financial institutions in Kuwait to advertise its goods and services on the Web. The institution maintains its own Internet department, responsible for designing and managing the company's Web site and for regulating the bank's wide range of e.banking services.

The KFH was incorporated in 1977, and since its creation, it has been "a non-interest based financial institution" where "creative and progressive financial engineering . . . go hand in hand with Islamic principles" (*http://www.kfh.com;* all subsequent references are to the Web site unless otherwise indicated) One of the main principles of Islam that is employed is *murabah,* which is "a trade finance transaction that provides a deferred payment option, whereby KFH purchases an identified asset(s) and then sells it to a potential client on the basis of a cost-plus-profit principle on a deferred payment basis." The following types of commodities can be financed with the KFH through *murabah* transactions (credit sales): cars, both new and used, furniture, electrical and electronic devices (such as computers), construction materials, boats, and spare parts. Credit sales can be repaid in up to sixty monthly payments or more. The Web site informs potential credit purchasers about the kinds of documents needed to engage in a transaction with the KFH, as well as some narrative details about the kinds of clients the KFH seeks. *Murabah's* legitimacy as a sound, Shariah-based approach to financing without interest is explored on the Web site. The last word is "God legalizes selling, but prohibits usury." Other Shariah-guided financial transactions pursued by the KFH include *mudaraba,* a trustee financing contract, or the financing of entrepreneurial projects, whereby KFH is guaranteed in advance a predetermined percentage of the final product or profits; *ijarah,* or leasing; and *istisna'a,* preproduction financing, such as that used to support oil processing or mining, construction, or manufacturing.

The KFH also is in the real estate investment business, both local and global. The Web site illustrates this activity with descriptions of the Danah Realty Fund, which offers investors an opportunity to profit from pooled capital gained from the purchase, selling, and management of real estate abroad. Pictured on the site are two residential locations that were purchased and sold for profit by the Danah Fund, the "Carlyle Lake" apartments in Atlanta, Georgia, and the Stratford apartments in Cary, North Carolina. In the local environment, in addition to helping Kuwaitis finance cars and furniture, the KFH participates in buying, selling, and developing real estate. The company portrays itself as an intermediary between the public's need for housing and the government's inefficiency in meeting the growing demand. Many young, married Kuwaitis in their early twenties expect to be on the list for housing, guaranteed by the government, for ten or more years. As a result, many will try to finance the purchase of a house on their own. This attitude parallels, in a way, young Americans' attitudes toward retirement and the incompetence of the Social Security system. The KFH claims that "the main objectives of the local real estate department are to encourage growth and development in Kuwait by expanding existing built-up areas and to ease the Government's burden of providing housing to nationals, thereby creating a wide range of opportunities for our youth venturing into a new phase of their lives." This process of "relieving the government's burden" illustrates one of the ways in which Islamist institutions participate in Kuwaiti politics, beyond a concern for social and cultural practices. Perhaps keeping Islamist politics in check is the fact that the Kuwaiti government holds 49 percent of the company's assets. The role of the state in supporting the KFH, however, suggests the depth of commitment that some members of the ruling family have to the Islamist movement.

In all, the KFH Web site reveals not only a business-savvy investing environment created and maintained along Islamic principles but also a technologically savvy use of new media in order to spread the economic activities of buying, selling, and investing that are Shariah based. The site has a very professional design matrix, using advanced Java script, photographs, a personalized greeting from the general manager of the financing sector, and an endorsement by several prominent Kuwaiti sheikhs. According to the Website, these sheikhs have reviewed all of the KFH's activities and give the institution high marks for observing the Shariah in all of its dealings. These factors, legal, technological, and aesthetic, all combine to make this Web site a perfect symbol of the economic power and social capital possible when Shariah meets the Internet in Kuwait.

A Web Site for Islamic Medicine

Those unfamiliar with classical Islamic traditions may not be aware of the rich heritage of Islamic medicine. Given the resurgence of global public interest in "alternative" approaches to healing, the Islamic Organization for Medical Sciences (IOMS) will be an institution of interest to a global audience, both Muslim and non-Muslim. Realizing the power of the Internet to spread the word of Islamic medicine's healing powers, former Minister of Public Health and President of the IOMS, Dr. Abdul Rahman Abdulla al-Awadi established a Web site to inform a global and local public about the activities of the center. This site illustrates the interesting convergence of new technologies with classical Islamic practices that help shape Muslim activities in Kuwaiti cyberspace.

One can access the home page for the IOMS at *http://www.islamset.com.* The center was established in 1984, with a large donation of private funds by the Marzook family, in addition to an emiri decree (which translates into state endorsement and funds). The center provides treatment for a number of ailments by incorporating traditional Islamic medicinal practices into modern medical science. In addition, the center serves as an organization that facilitates the pooling of Islamic knowledge of medical science worldwide. It also facilitates global dialogue among physicians, both Muslim and non-Muslim, who are encouraged to share knowledge about research, ethics, and ideas for future development of medical science in line with Islamic principles. This dialogue is sustained by the Web site, as well as by international conferences and publications supported by the center.

The World Health Organization (WHO) selected the IOMS in Kuwait as a regional center for training and research. The WHO and IOMS partnership has produced a series of seminars, the seventh of which was held in Kuwait in March 2002, which addressed the "Impact of Globalization on Development and Health Care Services in Islamic Countries." The first seminar identified fifty-two medicinal plants, recognized in classical Islamic sources to be used in treating various ailments. The second seminar established an agreement to avoid the use of alcohol in medicines used to treat children and pregnant women as a first step toward total elimination of alcohol from all drugs used to treat Muslims. The third seminar was held in Turkey and discussed the Islamic contribution to Western medicine. Subsequent conferences have been held in Karachi and Istanbul and have addressed issues of Islam and preventative medicine and Islamic medical approaches to drug use, psychologically harmful substances, and smoking.

Although the IOMS was established in 1984, it was not until the late 1990s that the organization began developing a Web presence. In 1997, when I first began analyzing the IOMS Web site, there had been less than 1,000 visitors to the site, and much of the site was still under construction. At that time, the home page had a clickable icon for displaying text in English or Arabic, but the Arabic link was still under construction. By 2002, the Web site had delivered a significant amount of content, in both English and Arabic, including a mission statement, a series of downloadable publications offering perspectives on

Islamic medicine, and an overview of the themes and rec-
ommendations generated by the organization's seminars
and international conferences. According to an article
posted on the site, written by the director, Dr. Awadi,

> The idea of establishing the IOMS emerged as
> one of several other ideas put forward at a time
> when the Islamic World was busting with prepara-
> tion for celebrating the advent of the fifteenth Hijri
> century, and when the Islamic awakening was at
> its zenith [1984]. . . . Questions came up one after
> the other in a rapid succession: How can we best
> celebrate this great event? Can we, Muslims, make
> some prominent and quite distinguished contribu-
> tion? Where do Muslims stand in respect to world
> civilizations now, and what was their place in past
> civilizations? What is Islam's view concerning mod-
> ern medical innovations?

In addition to setting up a center for the explicit purpose of healing the sick, and commemorating Muhammad's journey from Mecca to Medinah, the IOMS was
established in order to educate the world community
about the achievements of Muslim scientists throughout
history and to combat the effects of Orientalism. Dr.
Awadi explains:

> There have been persistent attempts, by both
> East and West, to undermine Islam and Muslims
> and to deliberately raise doubts about its role. One
> of these attempts by the enemies of Islam is to
> claim that it is only part of an overall world civiliza-
> tion that flourished for a certain period of time
> then declined. This is a flagrant falsity meant to
> obliterate the truth; for Islam is an eternal mes-
> sage from Allah to the whole of the human kind to

guide then [*sic*] to the right track. It is not the result of human development as is the case with the
rest of civilization.

Thus the center was established in part to correct the
historical record of the character of Islamic religion and
civilization, an important part of which was Islamic medicine and its great contribution to the world's sciences. We
are told:

> For five centuries under the umbrella of Islam,
> sciences, and medicine in particular, flourished
> and Muslims prolifically produced what later be-
> came the spring from which Europe gulped during
> its renaissance. Names started to shine as brightly
> as the sun among which were those of al-Razi, Ibn
> Sina, Ibnil Bitar, [*sic*] Ibnul Nafees, and al-
> Zahrawi. It was their strong and unwavering faith
> that formed the solid foundation for their magnifi-
> cent edifice of wisdom and knowledge. Their works
> were translated into all the living languages at that
> time. Thus Europe began to feast upon them and
> thrived in the process. Unfortunately, however, it
> later denied Islam and the Muslims their due
> credit as European writings appeared void of any
> mention of indebtedness to Arab merits. What is
> more aggravating, these European writings con-
> tained such malicious attacks against the Arabs
> and Muslims in general that it becomes incumbent
> upon us to stand up to these attacks.

There is a spectacular photograph on the IOMS Web
site of the mosque located at the center, which "has mag-
nificent Islamic architecture with a dome inside which is
painted with golden Islamic designs." The mosque is three
stories, we are told at the site, and the latest technologies

are present, a "lift" to the second floor where ladies are to pray, whereas the third floor "is set aside for the air-conditioning equipment." On the first floor there is enough space for 1,500 men to pray, and spectacular carpets, matching the designs in the dome, grace the floor. The center, including the architecturally significant mosque, is located in Shuweikh, an industrial area, near the Shuweikh campus of Kuwait University. The center, symbolized by the large golden dome of the mosque, is a vision of purity and divinity in a zone of dirt, trucks, and exhaust. The fact that the mosque is a prominent feature of this Islamic medical center suggests that healing processes within the organization are both medical and spiritual.

There are fifteen kinds of illnesses that are treated at the center using Islamic principles, including allergies, sinusitis, bronchitis, arthritis, diabetes, urinary tract infections, ulcers, migraines, constipation, and liver problems. According to the center's Web site, "Islam has got the means for good health which exceeds any that may be found in any program, old or new, of these we may mention hygiene, cleanliness, sports, bathing, praying, fasting, etc. . . . which have been referred to in the Holy Quran." In treating these illnesses, "this center is the only one in the world that has combined methods of modern medicine with treatment by herbal drugs." In terms of the impact of these treatments on Kuwaiti society, the site tells us, "it is our pleasure to say that Allah has granted us success in relieving pain and curing thousands of citizens." Once again we see that when Shariah meets the Internet, the place of Islam in local and global life can be enhanced via technologies that were created with much different purposes in mind.

A Web Site for Conversion

"Welcome to Enlightenment" is the phrase that opens the Islam Presentation Committee (IPC) home page. So far, over 12,000 visitors have received the good messages of Islam put forth into cyberspace by the IPC, at least according to the counter at the site. Some of these visitors choose to sign the guestbook. A perusal of comments left by visitors yields common themes, including praise for the messages available on the site, and praise for the IPC's advanced use of the Internet for *da'wa.* One such visitor, Mohammed Jimeh, writes, "alhamdulila that we are in the Net. May Allah give us the ability to make proper use of it." Similarly, Abdulssalam Muhammad al-Rumi states in his guestbook comments, "Congratulations on your beautiful WWW pages. It's a pleasure to see *da'wa* being done over the network. I hope that many people will learn to appreciate Islam thanks to your efforts." The IPC home page is accessible at *http://www. islam.com.kw.* An article in the local press explains that IPC Internet Services:

> introduce Kuwaiti traditions, culture, and objective teachings of Islam through the Internet said Khaled Muhammad, head of the section of Media and Publications Relations. People around the world are watching the IPC home page. IPC is receiving many e-mails from different parts of the world, particularly from the United States and other Western countries asking questions about Islam, Arabic language, Kuwaiti traditions, punishment in Islam, honor crimes, [the] dietary system in Islam, and Islamic injunctions for different aspects of life. Apart from answering the questions and clarifying the misconceptions about Islam, IPC is also giving information about Arabic language and details regarding Kuwait in general.[12]

On the front page of the IPC Web site are links to seven specialized topics, including a twelve-page exploration of

Islam as a religion, narratives describing the conversion of new Muslims, a page called "enlightenment," with links to thirteen different topics including Jesus, the moral system of Islam, Mary (Jesus' mother), Islam and human rights, the concept of worship, the story of Cat Stevens's conversion, and an introduction to the Quran. Also featured on this site are links to full texts of the Quran and Hadith, descriptions of Islamic centers around the world, with links to many of their home pages, and a section on women and Islam. Finally, there are links to a translation vehicle (Arabic) and an Islamic bookstore.

The IPC uses the conversion story of Cat Stevens as a means of encouraging people to embrace Islam. Cat Stevens's message is to read the Quran. He explains:

Everything I do is for the pleasure of God and pray that you gain some inspirations from my experiences. Furthermore, I would like to stress that I did not come into contact with any Muslim before I embraced Islam. I read the Qu'ran first and realized that no person is perfect. Islam is perfect, and if we imitate the conduct of the Holy Prophet (*pbuh*), we will be successful.

What Cat Stevens finds most inspiring about the Quran is that:

on reading the Qur'an, I now realize that all Prophets sent by God brought the same message. Why then were the Jews and Christians different? I know now how the Jews did not accept Jesus as the Messiah and that they had changed God's word. Even the Christians misunderstand His Word and called Jesus the son of God. Everything made so much sense. This is the beauty of the Qur'an; it asks you to reflect and reason and not to

worship the sun or moon but the One who has created everything.

Assuming visitors to the IPC site are inspired to want
to read the Quran, a translated version is available. To
provide guidance to the uninitiated, the IPC Web site
explains:

> The reader of the Qur'an must realize the fact
> that it is a unique book, quite different from the
> books one usually reads. . . . It does not contain in-
> formation, ideas, and arguments about specific
> themes arranged in a literary order. That is why a
> stranger to the Qur'an, on his first approach to it,
> is baffled when he does not find the enunciation of
> its theme or its division into chapters and sections
> or separate treatment of different topics and sepa-
> rate instructions for different aspects of life
> arranged in a serial order. He finds that it deals
> with creeds, gives moral instructions, lays down
> laws, invites people to Islam, admonishes the dis-
> believers, draws lessons from historical events, ad-
> ministers warnings, and gives good tidings, all
> blended together in a beautiful manner.

An "enlightenment section" exists on the Web site for
those "strangers to the Qur'an" on their "first approach,"
whereby certain key topics are discussed, along with support and guidance from the Quranic text and the com-
mentaries. For example, in the Islam and Human Freedom section, we are told that the Quran has a litany of
teachings on basic human rights—right to life, right to
justice, right to protection from slander and ridicule, right
to place of residence, right to asylum, right to seek knowledge, and right to religious freedom. These basic rights
and their delineation in the Quran are illustrated in an

easily accessible fashion. For example, in terms of the right to protection from slander, Q 49:11–12 is quoted: "Oh you who have attained the faith! No man shall deride (other) men: it may well be that those (whom they deride) are better than themselves . . . and neither should you defame one another, or insult one another."

The IPC also provides twelve proofs that Muhammad (*pbuh*) is the true Prophet of God. Proofs include an analysis of Muhammad's illiteracy as evidence of the Quran's divine inspiration; the perfection of Muhammad's life, whereby "he was a perfect example of being upright, merciful, compassionate, truthful, brave, generous, distant from all evil character and ascetic in all worldly matters, while striving solely for the reward of the Hereafter"; and proof of the true prophethood of Muhammad, based upon the fact that so many people from all over the world have chosen to follow the teachings of the Prophet (*pbuh*), even leaving their original faith to follow Islam instead.

The IPC site uses the latest technologies and processes of scientific reasoning to convey Islam's message. Billing itself as an organization "devoted to inter-religious and cultural studies" rather than as an aggressive conversion team, the IPC has achieved great success in gently leading people to recognize their innermost Muslim consciousness and to thus embrace the faith. The organization also uses an active program of community outreach in Kuwait, targeting mostly expatriate workers, across the social spectrum. For diplomatic communities and other white-collar guest workers in Kuwait, the IPC establishes social gatherings and exchanges where Muslims can meet with non-Muslims of a similar social class. A women's tea, for example, might including diplomats, diplomats' wives, and Kuwaiti women from the upper class. These engagements are especially common when embassies are hosting VIPs or other important guests, or during key Muslim holidays, such as Ramadan or Eid al-Fitr. The public affairs officers of the U.S. Embassy suggested that I visit the IPC in order to grasp something about Islam in Kuwait, demonstrating the close connections between them.

For the working class, the IPC hosts discussions with doctors or government officials where the topic may be the importance of good hygiene, lectures on how to properly handle perishable foods, or Muslim teachings on public health issues. For example, on April 1, 2002, the IPC gave a lecture for new converts on food and health in Islam. The lecturer, Dr. Hani al-Mazidi, from the Kuwait Institute for Scientific Research, stressed "the need of proper hygiene in food preparation and in cooking." He also "urged attendees to avoid junk food and fast food that are detrimental to human health" and observed that "every Muslim should wash his hands before eating and never eat or drink in broken wares, because some [of] its small pieces might get mixed up with the food, hence poisoning the body."[13] The IPC estimates that it converted approximately 2,000 people to Islam in 2002.[14]

The IPC headquarters is located in downtown Kuwait City, one block from the Sheraton Hotel and the first ring road. At the complex one will find a mosque, a bookstore, meeting rooms, and administrative offices. The director general of the organization is Abdul Wahab al-Shaya, a former pilot for the Kuwaiti Air Force and a member of one of Kuwait's key merchant families. In an interview with the director, I explained that I was studying how the Internet affects Kuwaiti society. And he stopped me, "You know we have our own Web site. We find these technologies quite well suited for spreading the message of Islam to the world." In our conversation, Abdul Wahab was clear to point out that the IPC had nothing to do with politics. He stressed that it was a religious organization designed to provide outreach to non-Muslims in Kuwait. "If the organization can succeed in leading one person to Islam then it has achieved its goal," he explained. With this said, Abdul Wahab asked me what religion I was and wanted to know if I had considered Islam. The rest of our conversation resembled the Web site, in that the director gave the basic proofs of the truth of Islam and asked me to consider them for my life. He was unable, or unwilling, to engage me on a social scientific level, and he suggested instead that I look within myself to seek the truth of God's plan for my life.

Our meeting concluded with a story, which emerged in response to a question of why and how the IPC was born. He explained that when he was a Kuwaiti Air Force pilot, he flew to the United States for some advanced tactical training at the U.S. Air Force Academy. On the plane, a female flight attendant began confiding in him that she had always been curious about Islam. This comment came in response to the fact that Abdul Wahab was reading the Quran. The flight attendant asked him questions about Islam with regard to what kinds of behaviors and attitudes were pleasing to Allah. She confided that she had been struggling with a loss of meaning in her life and was generally quite depressed. By the end of the flight, the attendant had been led to Islam by Abdul Wahab. He left her with a copy of the Quran and encouraged her to seek knowledge and guidance from the book. It was at that moment that Abdul Wahab realized what his true calling in life was—not flying planes but leading people to Islam. He learned from this experience that many people, in all walks of life, are struggling with spiritual bankruptcy. If he could only get them the knowledge they sought by presenting them with the teachings of the Quran, then he could make the world a much more secure place. The Internet site maintained by the IPC was created to advance this lifelong dream of Abdul Wahab al-Shaya.

Conclusion

The headline read, "Muslim Clerics in Kuwait Divided on Divorce by Internet, Mobile Message." At issue is the question, "Can a man divorce his wife by sending her an SMS text message by mobile phone, or an e-mail?" The response by Khalid al-Mathkour, Kuwait's most respected religious authority, is yes, "a simple text message on the mobile phone is enough."[15] The same is true for an e-mail divorce request. The explanation provided is that "it is as valid as a husband divorcing his wife in a face-to-face encounter, under Shariah laws allowing a man to divorce his wife by repeating the phrase 'I divorce you' three times." One of the conditions that must be met is that "the wife must check that the message was not sent as a prank by a third party."

This story epitomizes the main themes of this chapter: first, that Muslims are actively incorporating advanced communications technologies into their daily lives; second, that the presence of such new technologies produces new questions about religious observance and conduct; and third, that Islamists in Kuwait are relatively liberal in their attitudes toward both the utility and adaptability of new communications technologies to Kuwaiti social life. Just because Kuwait is a country and a society increasingly guided by Islamic principles does not mean that it will be any less high tech. On the contrary, a symbiotic relationship exists between religion and technology, as suggested by the earlier example, and by the analysis of Islamist Web sites in this chapter. Whereas Western social science and American policy makers often assume a symbiotic relationship between the Internet and democratization, and the Internet and western forms of social and economic liberalization, such assumptions are not realized in Kuwait, where local cultural values and politics are aided by global technologies to enhance the place of Islamic conservatism in everyday life.

Conclusion: Technological and Epistemological Challenges in Internet Studies—Lessons from an Ethnographic Approach

> Nothing is more difficult to arrange, more dangerous to manage or less certain to succeed than the introduction of new things.
> —Niccolo Machiavelli, *Il Principe*

The following pages attempt to pull together the main findings and contributions of this book to an understanding of the Internet in context. The conclusions deal with both the meaning of the ethnography for an understanding of Kuwaiti Internet practices and more general conclusions regarding epistemological challenges in Internet Studies. Throughout the course of the ethnography, this book illustrates that the context of Internet use matters significantly in the digital age, that online behavior is in part shaped by off-line variables such as power and political norms, cultural values such as gender relations, religion, and demographic factors such as age, socioeconomic status, and geographic location. This book provides a view of the Internet as a social construct, an extension of culture and human relationships in a non-Western context. Through an analysis of how the Internet fits into everyday life in an Islamic country, this book attempts to push Internet studies over the borders of Western centrism. It shows how the Internet merges with and transforms (and is transformed by) local communication patterns and human imaginations in Kuwait.

The Kuwaiti case teaches us that the synthesis of local culture with new technologies produces a social construction of the Internet reflective of that space. We see the ability of the Internet to morph in line with contextual variables in Islamist uses of the Internet to promote religious conservatism; in uses of the Internet by youths in ways both compatible and incompatible with Muslim family values; and in Kuwaiti women's hesitancy to use the Internet to oppose patriarchal power relationships, yet their interest in transgressing gender lines via the Internet. The ethnography suggests that we should soften our expectations for the rapid re-engineering of social and political spaces in the information age. Carolyn Marvin eloquently summarizes the point made here when she remarks that "electronic and other media precipitate new kinds of social encounters long before their incarnation in fixed institutional form."[1] In the Kuwaiti case, this observation helps explain why it is difficult to exactly pinpoint in a general way the meaning of the Internet beyond the narratives and online behavior of users. We are seeing small signs of new social relationships emerging in Kuwait, such as cyberdating, Islamist Web sites, and opportunities for women to voice their opinions to new, gender-mixed cyber publics. Yet these small shifts in everyday life have yet to be translated into formal institutionalizations of new power relationships, whether those governing gender relationships, the economy, or the political processes. The Internet has existed in Kuwait for more than a decade, but it was not until 2005 that women gained the vote, or were allowed to hold public office, youths are still living their lives in careful deference to social and parental expectations, and the economy is still largely state owned and petroleum based. The way in which Kuwaitis are able to tailor the Internet to their own needs illustrates the results of use, and the context of use rather than blunt technological determinism that promises to reshape local cultural landscapes in the Internet age. This finding encourages researchers to widen the focus of their gaze at the Internet's many meanings worldwide. Moreover, this study of the Internet in context also highlights a number of epistemological challenges in Internet Studies, as considered next.

Time and Transformation

In terms of epistemological challenges in Internet Studies, this book highlights several. First is the issue of time. Part of what makes study of the Internet difficult is that the technology is still in the early stages of its evolution. It is a relatively new thing, difficult to manage, and potentially unable to live up to all of the expectations attached to it. Drawing inspiration from Machiavelli's observation at the opening of this chapter, venturing out into the wilds of Internet Studies can be intellectually dangerous. The pace of change in Internet technologies, applications, and uses is so breakneck fast that concluding anything lasting about the technology's meaning is a real challenge. The rate of obsolescence equals the speed of innovation. For example, when I first began writing this book seven years ago, I opened the first chapter with the following observation:

When the Internet burst on the world political scene, analysts of this technology's future impact had a field day. New networked communications were going to change the face of international relations, the tools of war, the sovereignty of nation states, the global economy, the authenticity of local culture, and more. No longer would seemingly marginal voices, the rumblings of villages, the

whisperings of women in kitchens, the wretched
cries of the earth—go unheard. New media like the
Internet would change the power and place of
"voice," broadly defined.[2]

There is a reason that seven years later, this passage
is in the back of the book in a section designed to illustrate that the speed of obsolescence equals the rate of innovation. My intent when I first wrote these sentences
was to exaggerate, to capture the spirit of the wild predictions attached to the Internet by the early adopters and
interpreters of the technology. But what was once intended as absurd has become, in hindsight, a mere statement of fact. In less than a decade, the Internet *has* enabled many of the processes highlighted in the passage.

For example, because of the Internet, we live in a
global community of increasingly networked relations,
where boundaries between individuals are shrinking, and
lines between states are frequently interpenetrated by
flows of capital, data, and bodies, thus transforming the
location and meaning of "international" relations. The Internet has provided a forum for the marginal to take center stage, from the al-Qa'eda movement to gay Muslim
movements to the global WTO opposition, thus restructuring the nature of power and voice in the global community. The tools of war are changing before our eyes.
For example, in the Internet age, U.S. Secretary of Defense Donald Rumsfeld can correspond with Iraqi gener-
als by e-mail, asking about the possibilities of surrender
in the middle of a battle (the second Gulf War). Tightly co-
ordinated intelligence comes from hacking phone lines.
Information warfare, killer viruses, and the promise of
nanotechnology represent the changing face of war.

At the same time, the global economy has been transformed into an information marketplace, where intellectual capital is the most valuable commodity, bought and

sold for billions, from drugs to advice to distractions (such as yoga, computer games, aromatherapy). Entrepreneurial spirit drives the new economy and makes the boundaries of the market eternally fuzzy. The more the knowledge economy becomes entrenched, the greater the social, economic, and political divisions between industrialized, newly industrialized, and postindustrialized nations become. The authenticity of local culture becomes endangered, as new paradigms for wealth creation and human development are spun. Being competitive in the global economy, according to the World Bank, the International Monetary Fund (IMF), the G9, the ITU, the UNDP, USAID, the OECD, and the European Union (EU) means that a country has to have transparency in political and economic transactions, has to have elaborate credit regimes, a stock exchange, and privatization. Connectivity to communications links and a population trained to know what to do with IT also are essential to surviving economically in the twenty-first century. At the same time a country has to have political and economic stability and high ratings on human development indicators to attract foreign investment, vital to sustained economic growth. The parameters designed by the world community to lessen the gaps between rich and poor are creating a kind of economic McDonaldization, an attempt at one-size-fits-all patterns of development. A slogan heard repeatedly on CNN International for Intercontinental Hotels observes in a way symbolic of increasing global economic uniformity, "We're all the same, all over the world, one world, one hotel, Intercontinental." But as this book demonstrates, we are not all the same. We do not all find the same utility in the tools provided to us. We all lead different lives, have different dreams and desires, different needs and problems, and different ways of solving problems that arise.

The short time in which exaggeration becomes matter

of fact in Internet Studies presents an epistemological challenge for scholars in this field. The effect of "time" in Internet Studies is aptly captured by Neil Barrett in his *The State of the Cybernation*. He observes:

> Writing about the Internet is a bit like trying to shoot a speeding bullet with a bow and arrow. Even as your fingers hit the keyboard, new developments occur by the minute. The Net is an evolutionary beast, moving at an awesome pace, creating new opportunities and challenges for all of us."[3]

Books about the Internet are like freeze-frame shots of an object traveling at the speed of light. In some cases, the representation is just a blur of its true form; in others, clarity is just a distortion of a process moving too quickly to have a fixed form.

Chaos and Contradiction

The Internet is a complex set of relationships, tools, and contexts to the degree that it is a "nonlinear" technology. It is not a phenomenon whereby "simply adding up all the actions and intensions of its individual parts" one can understand the technology and its global and local meanings.[4] Rather, the Internet is chaotic. It is everywhere and nowhere, all at once, like Foucault's idea of power. It means different things to different people at different times of the day and in different circumstances. The technology is flexible enough to fill many different needs and purposes for the same user. In the words of one observer, "The Web is like an economy, in that the precise knowledge about the plans and strategies of individuals in the market would not suffice to understand the behavior of a market."[5] In other words, the whole of the Internet is more than the complex sum of its parts.

A number of choices are available to researchers in the emerging field of Internet Studies with which to confront the chaotic nature of the subject matter. Some, like Bernardo Huberman, choose to adapt statistical methods similar to the ones used by physicists to study chaotic systems, to find what Huberman calls "the laws of the Web." Huberman's explanation for adopting these methods is the following:

> Since following a single individual in her surfing behavior on the Web will not predict much about surfing in general, or how congestion takes place on the Internet, or the commercial success of given businesses, we must abandon such individual knowledge and replace it with something more aggregate, the behavior of the system as a whole. . . . We developed methods for treating large distributed systems that are largely inspired by the success that physics has had with explaining the behavior of matter in terms of its constituents, such as atoms and molecules. These methods are statistical in nature.[6]

In Huberman's epistemological framework, the desire is to know and bring order to the system as a whole. But how can something without a fixed form be known in its entirety?

The ethnographic approach to Internet Studies, like the one used for this study of Kuwait, exists at the polar opposite of Huberman's choice. Ethnographers *do* follow the surfing behavior of individual users as well as focus on the ways in which the context of use helps shape on-line and off-line technological meanings. The goal is not to understand the "system" as a whole but rather to explore the meaning of the Internet for individuals and small communities of users. Ethnographers operate with an epistemological framework that assumes that the Internet's ability to adapt to an unpredictable variety of environments means that it is difficult to really know the net outside of the ideas and images of individual users.

Some have said that the Internet, although in a constant state of change, does have fixed expressions, like a photograph of real life. It is packet switching, traceable flows of data, pinging, IP addresses, domain names, trackable user practices and patterns, surveyable links, hits, saliencies, and practices, and predictabilities such as who is shopping online, who is online, and what people are doing online, in addition to the many textual expressions of cyber identity contained in home pages, links, chatrooms, e-mail, message boards and Listservs™, and newsgroups. And yet pick up any book on the Internet and your are likely to find some kind of apology for the contingent nature of the conclusions provided by this study.

The chaotic nature of the Internet—given its unpredictability and flexibility, the complexity of user behaviors, and the inconclusive link between human behaviors online and that in the real world—creates challenges for social scientists. John Naughton, in his *A Brief History of the Future*, captures the spirit of chaos behind Internet Studies when he observes:

> When people learned I was writing a book about the history of the Internet they were often incredulous, not just because they thought the network wasn't old enough to have a history, but because they regarded it as somehow absurd that one should try and pin down something so fluid . . . writing about the Net is like skating on quicksand.[7]

Similarly, Bard and Soderquist, in their book *Netocracy* highlight the relationship between chaos and knowledge in Internet Studies when they observe: "The real challenge for social scientists or anyone trying to understand the Web is that a steady pattern will be difficult to fixate as new trends confront ever more violent countertrends."[8] The Internet reminds us, like no other technology "that change itself is the only thing that is permanent."[9] So this book, like all books in Internet Studies, provides still-life views of a highly dynamic phenomenon, and in the process it contributes to the history of a process globally significant, locally nuanced, and subject to radical fluctuations.

Global Reach, Local Innovation

Part of what makes the Internet increasingly challenging epistemologically is that more and more people around the world—with highly divergent cultures, identities, value systems, and interests—are becoming users. Haythornthwaite and Wellman have argued: "We are moving from a world of Internet wizards to a world of ordinary people routinely using the Internet as an embedded part of their lives."[10] As access has diffused, the characteristics of users also have evolved. Whereas early adopters tended to be "white, young, educated, North American men"[11] in the twenty-first century, more Internet users are women (more than 50% of Internet users in North America), non-American (64% of users live outside of North America), and older (e.g., number-1 hobby of British pensioners) than ever before.[12] With more of the general public using the Internet, the possibilities for more robust, yet chaotic, evidence of lifestyle change and new ways of acting socially, politically, and economically also emerge. We are given more over which to wonder, as we think about the increasing presence of information, the network, and more generalized human-machine interactions in our daily lives.

Conclusion

I hope this book has captured some of the spirit of wonderment that went into its construction. As I pound out these last sentences on my laptop, I am humbled by the speed with which my conclusions will be obsolete and inspired by the hope that my observations will draw attention to the scope, meaning, and purpose of the Internet as a global phenomenon. If this book stimulates future looks at the Internet in Kuwait, the Islamic world, and the world as a whole, then it will have achieved its goal. If it leads to better methodologies and sounder epistemologies with which to explore and explain processes of rapid technological change, then the exhaustion I now feel will have all been worth it.

Notes

Chapter 1

1. Plato, *The Phaedrus*, Translated by Benjamin Jowett, 275 (http://classics.mit.edu/Plato/phaedrus.html.) Accessed Feb. 22, 2003.

2. Harrison "Lee" Rainie, "Forward," in *Society Online: The Internet in Context*, ed. Philip N. Howard and Steve Jones (Thousand Oaks, Calif.: Sage, 2004), xi. For a more visual treatment of the history of the Internet see the PBS video, *Nerds 2.0*, which contains interviews with the early developers of the Internet as well as video footage of the actual machines upon which the first file transfers took place.

3. Robert Hobbes, *Hobbes' Internet Time Line*, 24. *(http://www.zakon.org/robert/internet/timeline)*.

4. See, for example, Barry Wellman and Caroline Haythornthwaite, eds., *The Internet and Everyday Life* (Oxford: Blackwell, 2002); Steve Woolgar, ed., *Virtual Society: Technology, Cyberbole, Reality,* (Oxford: Oxford University Press, 2002), 23–40; Bernardo A. Huberman, *The Laws of the Web: Patterns in the Ecology of Information* (Cambridge: MIT Press, 2001); Charles Ess and Fay Sudweeks, eds., *Culture, Technology, Communication: Towards an Intercultural Global Village* (Albany, New York: State University of New York Press, 2001); Daniel Miller and Don Slater, *The Internet: An Ethnographic Approach* (Oxford: Berg, 2001); Christine Hein, *Virtual Ethnography* (Thousand Oaks, Calif.: Sage, 2000).

5. See, for example, the Pew Charitable Trust Internet and the American Life project (http://www.pewinternet.org) and

Alan Neustadlt, John P. Robinson, and Meyer Kestnbaum, "Doing Social Research Online," in Barry Wellman & Caroline Haythornthwaite, eds., *The Internet and Everyday Life*, (Oxford: Blackwell, 2002), 186–211.

6. See Deborah L. Wheeler, "Blessings and Curses: Women and the Internet in the Arab World," in *Women and the Media in the Middle East*, ed. *Naomi Sakr* (London: IB Tauris, 2004), 138–168; Jo Tacchi, Don Slater, and Peter Lewis, "Evaluating Community Based Media Initiatives: An Ethnographic Action Research Approach," paper presented at the IT4D Conference, Wolfson College, Oxford University, July, 18th 2003; Miller and Slater, *The Internet: An Ethnographic Approach;* and Hein, *Virtual Ethnography.*

7. Examples include Miller and Slater, *The Internet*; Shanthi Kalathil and Taylor C. Boas, *Open Networks, Closed Regimes: The Impact of the Internet on Authoritarian Rule* (Washington, D.C.: Brookings Institution Press, 2003); Marcus Franda, *Launching into Cyberspace: Internet Development in Five World Regions* (Boulder, Colo.: Lynne Rienner, 2002); Eric Goldstein, *The Internet in the Middle East and North Africa: Free Expression and Censorship* (New York: Human Rights Watch, 1999); Nina Hachigian, "The Internet and Power in One-Party East Asian States," *Washington Quarterly* 25:3 (2002): 41–58.

8. Miller and Slater, *The Internet*, 1.

9. Wenhong Chen, Jeffrey Boase, and Barry Wellman, "The Global Villagers: Comparing Internet Use around the World," in *The Internet and Everyday Life*, 75.

10. Manuel Castells, *The Internet Galaxy* (Oxford: Oxford University Press, 2002), 10.

11. Castells, *The Internet Galaxy*, 20.

12. Castells, *The Internet Galaxy*, 20.

13. Vannevar Bush, "As We May Think," *Atlantic Monthly* (July 1945): 10.

14. Joseph C. R. Licklider, "Man–Computer Symbiosis," quoted in Janet Abbate, *Inventing the Internet* (Cambridge: MIT Press, 1999), 4.

15. Neil Randall, *The Soul of the Internet: Net Gods, Netizens*

and the Wiring of the World (London: International Thomson Computer Press, 1996), 353.

16. Ibid.

17. Abbate, *Inventing the Internet*, 2–3.

18. Abbate, *Inventing the Internet*, 3.

19. Abbate, Ibid.

20. Abbate, *Inventing the Internet*, 4.

21. Abbate, *Inventing the Internet*, 5.

22. Abbate, Ibid.

23. Abbate, *Inventing the Internet*, 2.

24. Hobbes, *Hobbes' Internet Time Line*, 32; in 1973, there were only thirty-five host computers linked to the ARPANET.

25. Wellman and Haythornthwaite, *The Internet and Everyday Life*, 19, quoting a study by UCLA Center for Communications Policy, 2000 "Surveying the Digital Future," http://www.ccp.ucla.edu.

26. J. E. Katz and P. Aspden, "A Nation of Strangers," *Communications of the ACM* 40:12 (1997): 81–86.

27. "Surveying the Digital Future," quoted in Wellman and Haythornthwaite, *The Internet and Everyday Life*, 19.

28. Quoted in Robert Putnam, *Bowling Alone* (New York: St. Martin's Press, 2000), 171.

29. John Perry Barlow, "Is There a There in Cyberspace?" *Utne Reader* (March, 1995), 50.

30. Woolgar, *Virtual Society*.

31. Katie Hafner and Mathew Lyon, *Where Wizards Stay Up Late* (New York: Touchstone, 1998), 3.

32. Jessica Mathews, "Forward," in Kalathil and Boas, *Open Networks, Closed Regimes*, ix.

33. Hafner and Lyon, *Where Wizards Stay Up Late*, 3.

34. Castells, *The Internet Galaxy*, 1.

35. Castells, Ibid.

36. Frank Webster, ed., "A New Politics?", in *Culture and Politics of the Information Age* (London: Routledge, 2001), 1.

37. Webster, *Culture and Politics*, 2.

38. The Mosaic Group, "Global Diffusion of the Internet Project: An Initial Inductive Study," 1 (http://mosaic.unomaha.edu/GDI1998/GDI1998.html).

39. See, for example, Pippa Norris, *The Digital Divide: Civic Engagement, Information Poverty, and the Internet Worldwide* (New York: Cambridge University Press, 2001); Mark Warschauer, *Technology and Social Inclusion: Rethinking the Digital Divide* (Cambridge: MIT Press, 2003).

40. Alexander Bard and Jan Soderquist, *Netocracy* (London: Pearson Education, 2002), 11.

41. Steve Jones, "What Is New about New Media?," in *New Media and Society* 1:1 (1998): 1–19.

42. Randall, *The Soul of the Internet*, 354.

43. Castells, *The Internet Galaxy*, 28.

44. For example, see Brian Winston, *Media, Technology, and Society: A History, from the Telegraph to the Internet* (London: Routledge, 1998); Tom Standage, *The Victorian Internet: The Remarkable Story of the Telegraph and the 19th Century's On-Line Pioneers* (New York: Berkley Publications Group, 1999); Wade Roland, *Spirit of the Web: The Age of Information from the Telegraph to the Internet* (Toronto: Key Porter Books, 1999); Jon Agar, Sarah Green, and Penny Harvey, "Cotton to Computers: From Industrial to Information Revolutions," in *Virtual Society: Technology, Cyberbole, Reality*, ed. Steve Woolgar, 264–285.

45. See, for example, the growing body of Internet research gathered and published by the Cultural Attitudes towards Technology and Communication Conference, organized annually by Charles Ess and Fay Sudwecks.

46. Neil Postman, *Technopoly: The Surrender of Culture to Technology* (New York: Vintage, 1993), quoted in Bard and Soderquist, *Netocracy*, 17.

47. Ibid.

48. Gene I. Rochlin, *Trapped in the Net: The Unanticipated*

Consequences of Computerization (Princeton: Princeton University Press, 1997), 7.

49. Ilkka Toumi, *Networks of Innovation: Change and Meaning in the Age of the Internet* (Oxford: Oxford University Press, 2002), 42.

50. Castells, "Series Editor's Preface: The Internet and the Network Society," in *The Internet and Everyday Life*, xxxi.

51. Ibid.

52. Sally Wyatt, Graham Thomas, and Tiziana Terranova, "They Came, They Surfed, They Went Back to the Beach: Conceptualizing Use and Non-Use of the Internet," in *Virtual Society: Technology, Cyberbole, Reality*, 25.

53. Kalathil and Boas, *Open Networks, Closed Regimes*, 137.

54. Rainie, *Society Online*, xi.

55. Rainie, *Society Online*, xii–xiii.

56. Rainie, *Society Online*, xiii; for other large N studies on the Internet and its meanings see General Social Survey, University of Chicago (http://www.norc.uchicago.edu/projects/gensoc.asp); The Home Net Study (http://www.homenet.hcii.cs.cmu.edu); The Survey 2001 Project (http://www.survey2001.nationalgeographic.com); PoliticalWeb (http://www.politicalweb.info); the UCLA Center for Communication Policy (http://www.ccp.ucla.edu).

57. Jo Tacchi, Don Slater and Peter Lewis, "Evaluating Community Based Media Initiatives: An Ethnographic Action Research Approach."

58. Kalathil and Boas, *Open Networks, Closed Regimes*, 137.

59. Putnam, *Bowling Alone;* Manuel Castells, *The Rise of The Network Society* (Oxford: Blackwell, 2000); Stephan Coleman, "Can the New Media Invigorate Democracies?" *Political Quarterly* 70:1(1999): 16–22; Webster, *Culture and Politics.*

60. Chen, Boase, and Wellman, in *The Internet and Everyday Life*, 107.

61. Sonia Liff, Fred Steward, and Peter Watts, "New Public Places for Internet Access: Networks for Practice-Based Learning and Social Inclusion," in *Virtual Society: Technology, Cyberbole, Reality*, 83.

62. Chia Siow Yue and Jamus Jerome Lim, eds., *Information Technology in Asia: New Development Paradigms* (Singapore: Institute of Southeast Asian Studies, 2002).

63. "Young Children Find Pornography on the Net," http://www.nua.ie/surveys.

64. "Indian Children Use Cyber Cafes to Get Online." http://www.nua.ie/surveys.

65. "Arab World Set to Go Online," http://www.nua.ie/surveys.

66. Hein, *Virtual Ethnography*, 155.

67. Ibid.

Chapter 2

1. "PC Penetration vs. Internet User Penetration in GCC Countries," *Madar Research Journal: Knowledge, Economy and Research on the Middle East* 1 (October: 2002): p. 10.

2. Robyn Greenspan, "Web Continues to Spread," *Jupiter Research Cyber Atlas* (http://www.cyberatlas.internet.com/big_picture/geographics/print.)

3. http://www.cybergeography.org/atlas/geographic.html.

4. Miral Fahmy, "Arabs Jeopardize Economic Future by Lagging on IT," *Reuters*, May 20, 2000, p. 1 (http://www.arabia.com/article/0,1690,Business|2074,00.htm).

5. Sonia Liff, Fred Steward, and Peter Watts, "New Public Places for Intenet Access: Networks for Practice-Based Learning and Social Inclusion," in *Virtual Society: Technology, Cyberbole, Reality*, ed. Steve Woolgar (Oxford: Oxford University Press, 2002), p. 83.

6. World Economic Forum, *Arab World Competitiveness Report 2002–2003* (Oxford: Oxford University Press, 2003), 32.

7. "Pan-Arab Mobile Phone Subscribers Reach 30 Million Mark in 2003," *AME Info*, January 27, 2004, 1 (http://www.ameinfo.com).

8. *CIA World Factbook*, 1998 (http://www.cia.gov/cia/publications/factbook/ku.html).

9. *Arab Times*, March 15, 1997, p. 1.

10. Muhammed S. al- Khulaifi, "Tathir al-Internet fi Mujtama'a min Wajha Nathar Tolab al Marhala al-Jaamaiyyah fi Mumlika al Arabiah al-Saudia," paper presented at the Islamic Information Resources Conference, Riyhad, Saudi Arabia, December 1998, p. 3.

11. Anh Nga Longva, *Walls Built on Sand: Migration, Exclusion, and Society in Kuwait* (Boulder, Colo: Westview Press, 1997).

12. This is true for the period 1996–1998, when the author was conducting fieldwork in the region. By mid 2000, when the author again had regular access to the *Arab Times* in Cairo, the "Internet Weekender" section had been eliminated, and instead a new section, called "On-Line," much smaller in scale and appearing more sporadically, replaced it. By 2000, more general coverage and advertising for Internet-related subjects occurred regularly in the press, suggesting that public knowledge of the Internet had evolved to the point that it was a normal part of everyday life, not requiring a special topics approach to its coverage.

13. Brian D. Loader, "The Governance of Cyberspace: Politics, Technology and Global Restructuring," in *The Governance of Cyberspace*, ed. Brian D. Loader (London: Routledge, 1997), 4.

14. "Internet Weekender," *Arab Times*, September 4–5, 1997, p. 6.

15. Ibid.

16. Ibid.

17. Personal communication with Gulfnet employees at the Info-World '97 trade show, Mishrif Fairgrounds, Kuwait.

18. Fiona McDonald, "In for a Cuppa . . . Dot Com," *Arab Times*, March 20, pp. 1, 8.

19 "50 Firms Submit Bids to Provide Internet Service," *Arab Times*, September 2, 1997, p. 2.

20. "Businessman Pursues Open University," *Arab Times*, March 30, 1997, p. 1.

21. Ibid.

22. "Alcohol Aims at Internet, "*Arab Times*, March 8, 1997, 1; "Terrorists On-Line," *Arab Times* January 26, 1997, p. 1.

23. *Al-Watan*, September 17, 1997, p. 6.

24. *Al-Qabbas*, January 17, 1997, p. 10.

25. *Al-Qabbas*, June 21, 1997, 10.

26. *P C Middle East*, "Almost One Million Online in Arab Countries," August 5, 1999, 1 (http://www.ditnet.co.ae/it-news/newsaug99/newsaug5.html).

27. *Al-Watan*, September 17, 1997, p. 7.

28. Rida al-Fili, "al-Thowra al-Technologiyya" (The Technological Revolution), *al-Rai al-Am*, August 10, 1997, p. 13.

29 "Kuwaitiun wal-Kitabun!" (Kuwaitis and Books!) *Al-Rai al-Am*, 30 August, 1997, 13; *Kuwait Times*, January 19, 1997, p. 4.

30. *Al-Rai al-Am*, August 29, 1997, p. 12.

31. http://www.ditnet.co.ae/itnews/newsaug99/newsaug5_tables.htm. By 2005, estimates are that @ 20% of the population in Kuwait is on-line (www.internetworldstats.com).

32. The Mosaic Group, "Global Diffusion of the Internet Project: An Initial Inductive Study," p. 1 (http://www.mosaic.unomaha.edu/GDI1998/GDI1998.html).

33. Joseph S. Nye Jr. and William A. Owens, "America's Information Edge: The Nature of Power," *Foreign Affairs* (March/April 1996), vol. 75, no. 2: 23.

34. Sara Baase, *A Gift of Fire: Social, Legal, and Ethical Issues for Computers and the Internet* (Upper Saddle River, N.J.: Prentice Hall, 2003); Reginald Whitaker, *The End of Privacy: How Total Surveillance Is Becoming a Reality* (New York: New Press, 1999).

35. John Perry Barlow, "Selling Wine without Bottles: The Economy of Mind on the Global Net," *The Electronic Frontier Foundation Archive* (http://www.eff.org/pub/Misc/Publications/John_Perry_Barlow/idea_economy.article).

36. Charles Ess, "Cosmopolitanism or Cybercentrism: A Critical Examination of the Underlying Assumptions of the Electronic Global Village" (http://www.drury.edu/faculty/ess/papers/cybercentrism.html).

37. Norman J. Vig, "Technology, Philosophy and the State," in *Technology and Politics*, ed. Michael Craft and Norman J. Vig (Durham, N.C.: Duke University Press, 1988), p. 19.

38. Ibid.

39. Shanthi Kalathil and Taylor C. Boas, *Open Networks, Closed Regimes: The Impact of the Internet on Authoritarian Rule* (Washington: D.C.: Brookings Institution Press, 2003).

40 Jon Alterman, *New Media, New Politics? From Satellite Television to the Internet in the Arab World.* (Washington, D.C. Washington Institute for Near East Policy, 1998), 69.

41. Ibid.

42. Augustus R. Norton, "New Media, Civic Pluralism, and the Slowly Retreating State," in *New Media in the Muslim World: The Emerging Public Sphere*, ed. Dale F. Eickelman and Jon W. Anderson (Bloomington: Indiana University Press, 1999), 27.

43. Ibid.

44. Dale Eickelman, "Bin Laden, the 'Arab Street,' and the Middle East's Democracy Deficit." *Current History*, 101:651 (January 2002): 36.

45. Gene I. Rochlin, *Trapped in the Net: The Unanticipated Consequences of Computerization* (Princeton: Princeton University Press, 1997), 8.

46. Rochlin, *Trapped in the Net*, 9.

47. Ibid.

48. Ibid.

49. McConnell International, "Ready? Net. Go! Partnerships

Leading the Global Economy," 4 (http://www.mcconnellinternational.com/ereadiness/ereadinessreport2.htm).

50. http://www.primeminister.gr/infosoc/en-01-01.htm.

51. Deborah L. Wheeler, "Living At E.Speed: A Look at Egypt's E.Readiness," in *Economic Challenges and Opportunities in the MENA Countries*, ed. Imed Limam (Cairo: American University in Cairo Press, 2003), 129–57.

52. See for example the papers presented at the Economic and Social Commission for Western Asia, "Western Asia Preparatory Conference for the World Summit on the Information Society (WSIS) Beirut, February 4–6, 2003.

53. Laleh D. Ebrahimian, "Socioeconomic Development in Iran through Information and Communications Technology," *Middle East Journal* 57:1 (Winter 2003): p. 93.

54. Karla J. Cunningham, "Jordan's Information Revolution: Implications for Democracy," *Middle East Journal* 56:2 (Spring 2002): 242).

55. Cunningham, *Jordan's Information Revolution*, 241.

56. Toby E. Huff, "Globalization and the Internet: Comparing the Middle Eastern and Malaysian Experiences," *Middle East Journal* 55:3 (Summer 2001): 441.

57. Interview with Jonnie Brown, Department of Commerce representative, U.S. Embassy, Kuwait, July 28, 1997.

58. "Trade Watchdogs Bemoan Rampant Music and Software Piracy in Kuwait," *Jordan Times* March 3, 2004, p. 11.

59. Kenneth Neil Cukier, "Internet Governance and the Ancient Regime," *Swiss Political Science Review* 5:1 (1999): 132 (http://www.sprc.ch).

60. Rochlin, *Trapped in the Net*, 8.

Chapter 3

1. Ya'qub Yusuf al-Hijji, *The Art of Dhow-Building in Kuwait* (Kuwait: Center for Research and Studies on Kuwait, 2001), vii.

2. Michael Herb, *All in the Family: Absolutism, Revolution, and Democracy in the Middle Eastern Monarchies* (Albany: State University of New York Press, 1999), 69.

3. *Kuwait: A MEED Practical Guide* (London: Middle East Economic Digest Limited, 1985), 11.

4. R. M. Burrell, "al-Kuwayt," *Encyclopedia of Islam* (Leiden: E. J. Brill, 1979), 572.

5. Burrell, *Encyclopedia of Islam*, 572.

6. Jacqueline S. Ismael, *Kuwait: Dependency and Class in a Rentier State* (Gainesville: University Press of Florida, 1993), 17.

7. Ismael, *Kuwait*, 28.

8. Ghanim al-Najjar, "Challenges Facing Kuwaiti Democracy," *Middle East Journal* 54:2 (Spring 2000): 243.

9. Ismael, *Kuwait*, 36.

10. Herb, *All in the Family*, 69.

11. Ibid.

12. Jill Crystal, *Oil and Politics in the Gulf: Rulers and Merchants in Kuwait and Qatar* (Cambridge: Cambridge University Press, 1995), 9.

13. Interview with a member of the al-Esa family, December 20, 2003.

14. "Politics Takes Backseat in Kuwait Poll," *al-Ahram Weekly* 646 July 10–16, 2003): 1 (http://www.weekly.ahram.org.eg/2003/646/re2.htm).

15. Ismael, *Kuwait*, 82.

16. Ibid.

17. Ibid.

18. Ibid.

19. Mary Ann Tétreault, *Stories of Democracy: Politics and Society in Contemporary Kuwait* (New York: Columbia University Press, 2000) 59.

20. Ismael, *Kuwait*, 82.

21 al-Najjar, "Challenges," 245.

22. al-Najjar, "Challenges," 247.

23. Ibid.

24. Haya al-Mughni, "From Gender Equality to Female Subjugation: The Changing Agendas of Women's Groups in Kuwait," in *Organizing Women: Formal and Informal Women's Groups in the Middle East*, ed. Dawn Chatty and Annika Rabo (Oxford: Berg, 1997), 195.

25. al-Najjar, "Challenges," 252.

26. Ibid.

27. Tétreault, *Stories of Democracy*, 9.

28. Ibid. See also, Mary Ann Tétreault, "Civil Society in Kuwait: Protected Spaces and Women's Rights," *Middle East Journal* 47 (Spring 1993): 275–91.

29. For more on the struggle of human rights organizations in Kuwait, see Tétreault, *Stories of Democracy*, 199–205.

30. al-Najjar, "Challenges," 256.

31. Crystal, *Oil and Politics in the Gulf*, 1.

32. Muhammad Rumaihi, *al-Kuwait lysa al-Naftan* (Beirut: Dar al-Jadid, 1995); Rumaihi, *Beyond Oil: Unity and Development in the Gulf* (London: Al-Saqi Books, 1986).

33. Tétreault, *Stories of Democracy*, 220.

34. Ibid.

35. al-Najjar, "Challenges," 249.

36. "Communications in the State of Kuwait," a special supplement to *Communications MEA* (Herts: Informtion Technology, 1993), 14–18.

37. Jamal al-Menayes, "Television Viewing Patterns in the State of Kuwait after the Iraqi Invasion," *Gazette: A Journal of International Communications*, 56(1996): 124.

38. Survey results published in *The Star*, July 23, 1998 (http://www.star.arabia.com/980730/TE2.html).

39. Tétreault, *Stories of Democracy*, 76.

40. Ibid.

41. "Communications in the State of Kuwait," 12.

42. Claudia Farkas Rashoud, *Kuwait: Before and after the Storm* (Kuwait City: The Kuwait Bookshop, 1992), 106.

43. John Levins, *Days of Fear: The Inside Story of the Iraqi Invasion and Occupation of Kuwait* (Dubai: Motivate Publishing, 1997), 364.

44. Levins, *Days of Fear,* 365.

45. Personal communication, Dr. Ahmed al-Baghdadi, chairman of the Political Science Department, Kuwait University, May 1997.

46. Reporters without Borders, "Kuwait 2003 Annual Report," (http://www.rsf.org).

47. "Kuwait Closes al-Jazeera Office," *al-Bawaba.com,* November 4, 2002 (http://www.gvnews.net/html).

48. *UNESCO Yearbook* (New York: UNESCO, 1996), iv–120.

49. Fayad E. Kazan, "Kuwait," in *Mass Media in the Middle East: A Comprehensive Handbook,* ed. Yahya R. Kamalipour and Hamid Mowlana (Westport, Conn: Greenwood Press, 1994), 147.

50. Wendy Kristianasen, "Kuwait's Islamists, Officially Unofficial," *Le Monde Diplomatique,* June 4, 2002, p. 2 (http://www.mondediplo.com/2002/06/04kuwait).

51. Shafiq al-Ghabra, "Balancing State and Society: The Islamic Movement in Kuwait," *Middle East Policy* 5:2(May 1997): 59.

52. Shafiq al-Ghabra, "Kuwait and the Dynamics of Socio-Economic Change," *Middle East Journal* 51:3(Summer 1997): 367.

53. al-Ghabra, "Balancing State and Society," 60.

54. al-Ghabra, "Balancing State and Society," 61.

55. al-Ghabra, "Balancing State and Society," 63

56. Tétreault, *Stories of Democracy,* 150.

57. Tétreault, *Stories of Democracy,* 177–80.

58. Khalil Osman, "Kuwaiti 'Royal Family' Trying to Mend a Bursting Dam with Sticky Tape," *Crescent International* (February 2001): 3 (http://www.muslimedia.com/arcives/oaw01/kuw-burst.htm).

59. Osman, "Kuwaiti . . . ," 3.

60. Osman, "Kuwaiti . . . ," 2.

61. al-Ghabra, "Balancing State and Society," 62.

62. *Kuwait Times,* January 9, 2002, p. 1.

63. Associated Press, December 1, 2001, p. 2.

64. Ibid.

65. Ibid.

66. Associated Press, December 1, 2001, p. 4.

67. *Kuwait Times,* January 9, 2002, p. 2.

68. Ilene R. Prusher, "Rise of Islamists Veils Liberalism," *Christian Science Monitor,* April 24, 2000, p. 1 (http://www.csmonitor.com/durable).

69. Kristianasen, "Kuwait's Islamists," 3.

70. Kelly Machinchick, "Kuwaiti Women's Suffrage Movement Comes to Washington," *Washington File,* February 11, 2003, p. 1. Note: In May 2005 Kuwaiti women were awarded full political rights.

71. Mary Ann Tétreault," Frankenstein's Lament in Kuwait," *Foreign Policy in Focus,* November 29, 2001, p. 3 (http://www.fpif.org/commentary/2001/0111kuwait.body.html).

72. Jim Landers, "Separate but Equal?," *Dallas Morning News,* March 9, 2003, p. 2J.

73. "The Kuwaiti Way of Life," *Embassy of the State of Kuwait* (2000), p. 1 (http://www.kuwait-embassy.org).

74. Haya al-Mughni, *Women in Kuwait: The Politics of Gender* (London: Saqi Books, 1993), 148.

75. Dr. Yusif al-Qaradawi, *Makanah al-Marah fi Islam* (The Status of Women in Islam) (Cairo: Islamic Home Publishing Co., 1997), 67.

76. "Kuwaiti Women Strongly Resent Idea of Staying at Home," *Arab Times*, January 8, 1994, p. 1.

77. A comment made during a conversation with a senior female member of the American diplomatic mission in Kuwait.

78. See, for example, *al-Samra (February 1997)*; Muntada al-Marah wa Sana' al-Qirar: Bahath wa-Awraq al-Amal (Kuwait: Women's Cultural and Social Society, 1996), 61–73.

79. Interview with a working woman at the Ministry of Education, August 6, 1997.

80. Al-Mughni, "From Gender Equality to Female Subjugation," 195.

81. Interview with Badria al-Awadi, Kuwait City, July 15, 1997.

82 John West, "Kuwait Bikers Seek Thrills Despite Spills," *Reuters* August 24, 2003, p. 1 (http://www.aduni.org/~john-west).

83. Ibid.

84. Ibid.

85. Ragab el-Damanhor, "Kuwaiti Youths Encouraged Not to Miss Prayers," *Islam on Line*, October 7, 2003, p. 1 (http://www.islamonline.net).

86. Ibid.

87. al-Mughni, *Women in Kuwait*, 50.

88. Ibid.

89. al-Mughni, *Women in Kuwait*, 53.

90. al-Ghabra, "State and Society in Balance," 59.

91. Interview with liberal Kuwaiti female activist, February 8, 1997, Kuwait City.

92. Interview with liberal Kuwaiti female activist March 20, 1997, Kuwait City.

93. Interview with Yusif al-Ibrahim at Kuwait University, January 13, 1997.

94. West, *Kuwait Bikers*, 2.

Chapter 4

1. Dale Spender, *Nattering on the Net* (Toronto: Garamond Press, 1995), xvi.

2. *Jordan Times*, November 5, 2002, p. 1.

3. Najat Rochdi, "Cultural Boundaries and Cyberspace," *Women's Wire* (2002), p. 1 (http://www.womenswire.org/opinion.htm).

4. Amal Bouhabib, "Wooing Women onto the World Wide Web," *Daily Star* (http://www3.estart.com/arab/women/www.html) accessed January 28, 2003.

5. Nancy Hafkin and Nancy Taggart, *Gender, Information Technology, and Developing Countries: An Analytic Study* (Washington, D.C.: Academy for Educational Development/ USAID, 2001), 16.

6. Olivia Acosta and Mari Hartl, "Women and the Information Revolution," *Women 2000* 1 (October 1996): 4 (New York: United Nations Division for the Advancement of Women) (http://www.un.org/dpcsd/daw).

7. Note: The names of the women interviewed here have been changed to protect their privacy.

8. Harriet Beecher Stowe, *Pink and White Tyranny* (New York: Macmillan, 1989).

9. If one looks at the employment classified ads, those calling for high-tech jobs rarely specify job qualifications along gender lines. In contrast, jobs that require driving and selling or that are management related are likely to use gender-specific language in their advertisements for the position.

10. Linda Low, "Social and Economic Issues in an Information Society: A Southeast Asian Perspective," *Asian Journal of Communications* 6:1 (1996): 12.

11. Yusif al-Qinanie, *Pages from Kuwait's History* (Arabic) (Kuwait: Government Printing House, 1968), 66, quoted in Haya al-Mughni, *Women in Kuwait: The Politics of Gender* (London: Saqi Books, 1993), 42.

12. When pressed on this issue, several women responded

that in Kuwait, the women who needed to be reached the most with regard to their rights were not computer literate, thus the Internet would do little to help them. On the contrary, hotlines for domestic abuse, special counseling offices for dealing with marital problems, social agencies dedicated to dealing with post-traumatic stress problems created by the terrors of the Iraqi invasion, and pamphlets printed in simple Arabic explaining women's legal rights under the Sharia (e.g., the fact that women can write into the prenuptial agreement that a man cannot take another wife) are the best channels of communication to use in campaigns for increasing women's access to social justice in Kuwait, according to a composite of women's rights advocates I interviewed.

13. Carolyn Marvin, *When Old Technologies Were New: Thinking about Electric Communication in the Late Nineteenth Century* (Oxford: Oxford University Press, 1988), 5.

14. Edward Talero and Phillip Gaudette, "Harnessing Information for Development: A Proposal for a World Bank Group Strategy," *World Bank* (March 1995): 2 (http://www.world-bank.org/html/fpd/harnessing).

Chapter 5

1. Elizabeth Warnock Fernea, ed., *Childhood in the Middle East* (Austin: University of Texas Press, 1995), xiii.

2. Saif Abbas, unpublished survey, Kuwait University Student Attitudes toward the Internet, 1997; Moosa L. al-Mazeedi and Ibrahim A. Ismail, "The Educational and Social Effects of the Internet on Kuwait University Students," paper presented at the Conference on Information Superhighway, Safat, Kuwait, March 3–5, 1998; Hassan Abbas, "Internet's Impact on Kuwaiti Youth," *Kuwait Times*, October, 14, 2001, p. 1.

3. Hassan Abbas, "Internet's Impact," p. 1.

4. DITnet, "Internet Reaches the Layman in the Middle East," August 5, 1999, 1 (http://www.ditnet.co.ae/itnews/me_internet/ecomprofiles.html).

5. Mazeedi and Ismail, "The Educational . . . ," 4.

6. Mazeedi and Ismail, "The Educational . . . ," 6.

7 Hassan Abbas, "Internet's Impact," 1.

8. Kenneth Neil Cukier, "Internet Governance and the Ancient Regime," *Swiss Political Science Review* 5:1, (1999): 133.

9. "Kuwait to Add Internet to Government Schools," *Kuwait Times,* October 13, 2001, p. 1.

10. Deborah Wheeler, "The Internet and Public Culture in Kuwait," *Gazette* 63:2–3 (2001): 198.

11. Interview with Professor Robert Sulayman Bower, American University, Cairo, December 10, 2001.

12 Note: in 1997, Internet cafes charged 3KD per hour, about ten dollars. Sa'ad's observation that Internet access in cafés is $1.50 per hour represents a significant drop in the cost of access over a period of four years.

13. *Halal* means that food that is compatible with Islamic guidelines. In other words, products do not contain alcohol or pork products, and meat and chicken products are raised free range, fed a healthy diet, and slaughtered according to Islamic guidelines.

Chapter 6

1. *al-Watan,* January 29, 1997, p. 1.

2. Bobby S. Sayyid, *A Fundamental Fear: Eurocentrism and the Emergence of Islamism* (London: Zed Books, 1997), 1

3. *Seattle Times,* July 1, 2002, p. A3.

4. *Al-Sharq al-Awsat,* October 13, 2001, p. 1.

5. *Los Angeles Times,* November 11, 2001, p. 4.

6. Tétreault, *Stories of Democracy,* 211.

7. Ibid.

8. Ibid.

9. Ibid.

10. Tétreault, *Stories of Democracy*, 212.

11. Mark Poster, *What's the Matter with the Internet?* (Minneapolis: University of Minnesota Press, 2001), 176.

12. *Kuwait Times*, January 22, 1997, p. 20.

13. "IPC Holds Lecture for New Converts," *Kuwait Times*, April 1, 2002, p. 1.

14. "IPC Honors Converts," *Kuwait Times*, October 23, 2002, p. 1.

15. Agency France Presse, July 14, 2001, p. 1.

Conclusion

1. Carolyn Marvin, *When Old Technologies Were New: Thinking About Electric Communication in the Late Nineteenth Century* (Oxford: Oxford University Press, 1988), 3.

2. For samples from the variety of discourses that makes up this sense of awe at the power of new media in world affairs, including satellites and the Internet, see Douglas Kellner, *The Persian Gulf TV War* (Boulder, Colo.: Westview Press, 1992); Manuel De Landa, *War in the Age of Intelligent Machines* (New York: Zone Books, 1994); James F. Larson and Heung-Soo Park, *Global Television and the Politics of the Seoul Olympics* (Boulder, Colo.: Westview Press, 1993); Warren P. Strobel, *Late-Breaking Foreign Policy: The News Media's Influence on Peace Operations* (Washington, D.C.: U.S. Institute of Peace, 1997); Strobe Talbot, "Democracy and the National Interest," *Foreign Affairs* 75:6 (November–December 1996): 47–63; Philip M. Taylor, *War and the Media: Propaganda and Persuasion in the Gulf War*, 2nd ed. (Manchester: University of Manchester Press, 1998); *Taken by Storm: The Media, Public Opinion, and U.S. Foreign Policy in the Gulf War*, Ed. W. Lance Bennett and David Paletz (Chicago: University of Chicago Press, 1994).

3. Neil Barrett, *The State of the Cybernation: Cultural, Political and Economic Implications of the Internet* (London: Kogan Page, 1997), 7.

4. Bernardo A. Huberman, *The Laws of the Web: Patterns in the Ecology of Information* (Cambridge: MIT Press, 2001), 21.

5. Huberman, *The Laws of the Web*, 23.

6. Ibid.

7. John Naughton, *A Brief History of the Future: The Origins of the Internet* (Collingdale: Diane Publishing, 1999), 267.

8. Bard and Soderquist, *Netocracy*, xii.

9. Ibid.

10. Wellman and Haythornthwaite, *The Internet and Everyday Life*, 6.

11. Ibid.

12. http://www.nua.ie/surveys/index.

Bibliography

Abbas, Hassan. "Internet's Impact on Kuwaiti Youth." *Kuwait Times*, October, 14, 2001, p. 1.

Abbas, Saif. Unpublished survey. "Kuwait University Student Attitudes toward the Internet," 1997.

Abbate, Janet. *Inventing the Internet.* Cambridge: MIT Press, 1997.

Acosta, Olivia and Mari Hartl. "Women and the Information Revolution." *Women 2000* 1 (October 1996): 4. New York: United Nations Division for the Advancement of Women. http://www.un.org/dpcsd/daw.

Agar, Jon, Sarah Green, and Penny Harvey. "Cotton to Computers: From Industrial to Information Revolutions." in *Virtual Society: Technology, Cyberbole, Reality,* edited by Steve Woolgar, 246–85. Oxford: Oxford University Press, 2002.

Agency France Press. July 14, 2001, p. 1.

Agre, Philip E. "The Internet and Public Discourse." *First Monday* 3:3:10–15, 1998.

"Alcohol Aims at Internet."1997. *Arab Times,* March 8, 1997, p. 1.

Alterman, Jon. *New Media, New Politics? From Satellite Television to the Internet in the Arab World.* Washington, D.C.: Washington Institute for Near East Policy, 1998.

Anderson, Jon and Dale Eickelman. "Convergence of Media Technologies in the Middle East." *Middle East Insight* (March–April 1999): 59–62.

"Arab World Set to Go On-line." http//www.nua.ie/surveys (accessed 2003).

Arthus-Bertrand, Yann. *Kuwait from Above.* Mansouria: Center for Research and Studies on Kuwait, 1998.

Aspray, William, and Bernard O. Williams. "Arming American Scientists: NSF and the Provision of Scientific Computing Facilities for Universities, 1950–1973. *Annals of the History of Computing* 16:4 (1994): 60–74.

Associated Press. December 1, 2001, p. 2.

Baase, Sara. *A Gift of Fire: Social, Legal and Ethical Issues for Computers and the Internet.* Upper Saddle River, N.J.: Prentice Hall, 2003.

al-Baghdadi, Ahmad. "Fi Mafhum al-Thaqafa wa al-Thaqafa al-Kuwaiti." *Alam al-Fikr* (June 1996): 9–22.

Baran, Paul. *On Distributed Communications.* 12 volumes. RAND Report series. Santa Monica, Calif.: RAND Corporation, 1964.

Bard, Alexander, and Jan Soderquist. *Netocracy.* London: Pearson Education, 2002.

Barlow, John Perry. "Is There a There in Cyberspace?" *Utne Reader (March 1995): 45–50.*

_____. "Selling Wine without Bottles: The Economy of Mind on the Global Net," The Electronic Frontier Foundation Archive. http://www.eff.org/pub/Misc/Publications/John_Perry_Barlow/idea_economy.article (accessed 1999).

Barrett, Neil. *The State of the Cybernation: Cultural, Political and Economic Implications of the Internet.* London: Kogan Page, 1997.

Bennett, Lance, and David Paletz, eds. *Taken by Storm: The Media, Public Opinion and U.S. Foreign Policy in the Gulf War.* Chicago: University of Chicago Press, 1994.

Bouhabib, Amal. "Wooing Women onto the World Wide Web." *Daily Star.* http://www3.estart.com/arab/women/www.html (accessed January 28, 2003).

Burkhart, Grey, and Sy Goodman. "The Internet Gains Acceptance in the Persian Gulf." *Communications of the ACM* 43:1 (March): 19–24.

Burrell, R. M. "al-Kuwayt." *Encyclopedia of Islam.* Leiden: E. J. Brill, 1979, 572–576.

Bush, Vannevar. "As We May Think." *Atlantic Monthly* (July 1945): 10.

"Businessman Pursues Open University." *Arab Times,* March 30, 1997, p. 1.

Carey, James W. "Mass Media and Democracy: Between the Modern and the Postmodern." *Journal of International Affairs* 47:1 (Summer 1993): 1–22.

Carlyle, Ralph Emmett. "Open Systems: What Price Freedom?" *Datamation 1* (June 1988): 54–60.

Castells, Manuel. *The Internet Galaxy.* Oxford: Oxford University Press, 2002.

_____. *The Rise of The Network Society.* Oxford: Blackwell, 2000.

Cerf, Vinton. "How the Internet Came to Be." In *The Online Users Encyclopedia,* edited by B. Aboba. New York: Addison Wesley, 1993.

Chen, Wenhong, Jeffrey Boase, and Barry Wellman. "The Global Villagers: Comparing Internet Use around the World." In *The Internet and Everyday Life,* edited by Barry Wellman and Caroline Haythornthwaite. Oxford: Blackwell, 2002.

Coleman, Stephan. "Can the New Media Invigorate Democracies?" *Political Quarterly* 70:1 (1999): 16–22.

"Communications in the State of Kuwait." A special supplement to *Communications MEA.* Herts: Information and Technology, 1993.

Crystal, Jill. *Oil and Politics in the Gulf: Rulers and Merchants in Kuwait and Qatar.* Cambridge: Cambridge University Press, 1995.

Cukier, Kenneth Neil. "Internet Governance and the Ancient Regime." *Swiss Political Science Review* 5:1 (1999): 132–57.

Cunningham, Robert B., and Yasin K. Sarayrah. *Wasta: The*

Hidden Force in Middle Eastern Society. London: Praeger, 1993.

el-Damanhor, Ragab. "Kuwaiti Youths Encouraged Not to Miss Prayers." *Islam on Line* October 7, 2003, p. 1. http://www.islamonline.net.

De Landa, Manuel. *War in the Age of Intelligent Machines*. New York: Zone Books, 1994.

DITnet. "Internet Reaches the Layman in the Middle East." http://www.ditnet.co.ae/itnews/me_internet/ecompro-files.html (accessed August 5, 1999).

Ebrahimian, Laleh D. "Socioeconomic Development in Iran through Information and Communications Technology," *Middle East Journal* 57:1 (Winter 2003): p. 93.

Edwards, Paul N. *The Closed World: Computers and the Politics of Discourse in Cold War America*. Cambridge: MIT Press, 1996.

Eickelman, Dale. "Bin Laden, the 'Arab Street' and the Middle East's Democracy Deficit." *Current History* 101:651 (January, 2002): 36–39.

Einstein, Elizabeth. *The Printing Press as an Agent of Change: Communications and Cultural Transformations in Early-Modern Europe*. 2 volumes. Cambridge: Cambridge University Press, 1979.

Ess, Charles. "Cosmopolitanism or Cybercentrism: A Critical Examination of the Underlying Assumptions of the Electronic Global Village." http://www.drury.edu/faculty/ess/papers/cybercentrism.html (accessed April 15, 2001).

Ess, Charles and Fay Sudweeks, eds. *Culture, Technology, Communication: Towards an Intercultural Global Village*. Albany: State University of New York Press, 2001.

Fahmy, Miral. "Arabs Jeopardize Economic Future by Lagging on IT." *Reuters*, May 20, 2002, p. 1. http://www.arabia.com/article/0,1690,Business|2074,00.htm.

Fernea, Elizabeth Warnock, ed. *Childhood in the Middle East*. Austin: University of Texas Press, 1995.

"50 Firms Submit Bids to Provide Internet Service." *Arab Times*, September 2, 1997, p. 2.

al-Fili, Rida. "al-Thowra al-Technologiyya" (The Technological Revolution). *Al-Rai al-Am*, August 10, 1997, p. 13.

Franda, Marcus. *Launching into Cyberspace: Internet Development in Five World Regions*. Boulder, Colo.: Lynne Rienner, 2002.

Gabbard, Brian C. and George S. Park. *The Information Revolution in the Arab World: Commercial, Cultural and Political Dimensions—The Middle East Meets the Internet*. Santa Monica: RAND Corporation, 1996.

al-Ghabra, Shafiq. "Balancing State and Society: The Islamic Movement in Kuwait." *Middle East Policy* 5:2 (May 1997): 58–72.

_____. "Kuwait and the Dynamics of Socio-Economic Change." *Middle East Journal* 51:3 (Summer 1997): 358–372.

Gher, Leo A., and Hussein Y. Amin, eds. *Civic Discourse and Digital Age Communications in the Middle East*. Stamford, Conn: Ablex, 2002.

Goldstein, Eric. *The Internet in the Middle East and North Africa: Free Expression and Censorship*. New York: Human Rights Watch, 1999.

Greenspan, Robyn. "Web Continues to Spread." *Jupiter Research Cyber Atlas*. http://www.cyberatlas.internet.com/big_picture/geographics/print (accessed Oct. 25, 2002).

Guazzone, Laura, ed. *The Islamist Dilemma: The Political Role of Islamist Movements in the Contemporary Arab World*. Reading, PA: Ithaca Press, 1995.

Hachigian, Nina. "The Internet and Power in One-Party East Asian States." *Washington Quarterly* 25:3 (2002): 41–58.

El-Hadary, Mohammed. "No Plans to Control Satellite." *Arab Times*, March, 25, 1997, p. 1.

Hafkin, Nancy and Nancy Taggart. *Gender, Information Technology, and Developing Countries: An Analytic Study*. Washington, D.C.: Academy for Educational Development/USAID, 2001.

Hafner, Katie, and Mathew Lyon. *Where Wizards Stay Up Late*. New York: Touchstone, 1998.

Hein, Christine. *Virtual Ethnography.* Thousand Oaks, Calif.: Sage, 2002.

Herb, Michael. *All in the Family: Absolutism, Revolution and Democracy in the Middle Eastern Monarchies.* Albany: State University of New York Press, 1999.

al-Hijji, Ya'qub Yusuf. *The Art of Dhow-building in Kuwait.* Kuwait: Center for Research and Studies on Kuwait, 2001.

Hobbes, Robert. *Hobbes' Internet Time Line.* http:www.zakon. org/robert/internet/timeline (accessed May 22, 2003).

Huberman, Bernardo A. *The Laws of the Web: Patterns in the Ecology of Information.* Cambridge: MIT Press, 2001.

Hughes, Thomas P. *Rescuing Prometheus.* New York: Pantheon, 1998.

Hussain, Fatima. "Kum alayna al-Nantathr." *Al-Samra* (February 1997): 16–18.

"Indian Children Use Cyber Cafes to Get On-line." http:// www.nua.ie/surveys (accessed Sept. 22, 2004).

"Internet Weekender." *Arab Times,* September 4–5, 1997, p. 6.

"IPC Holds Lecture for New Converts." *Kuwait Times,* April 1, 2002, p. 1.

"IPC Honors Converts." *Kuwait Times,* October 23, 2002. p. 1.

Ismael, Jacqueline S. *Kuwait: Dependency and Class in a Rentier State.* Gainesville: University Press of Florida, 1993.

Jones, Steve. "What is New about New Media?" *New Media and Society* 1:1: (1998): 1–19.

Jordan Times. November 5, 2002, p. 1.

Kahin, B., and J. Keller, eds. *Public Access to the Internet.* Cambridge: MIT Press, 1995.

Kalathil, Shanthi and Taylor C. Boas. *Open Networks, Closed Regimes: The Impact of the Internet on Authoritarian Rule.* Washington, D.C.: Brookings Institution Press, 2003.

Karl, Terry Lynn. *Paradox of Plenty: Oil Booms and Petrol States.* Berkeley: University of California Press, 1997.

_____. "The Perils of the Petrol State: Reflections on the Paradox of Plenty." *Journal of International Affairs* (Fall 1999) 53#1, 3148.

Katz, J. E., and P. Aspden. "A Nation of Strangers." *Communications of the ACM* 40:12 (1997): 81–6.

Kazan, Fayad. E. "Kuwait." In *Mass Media in the Middle East: A Comprehensive Handbook*, edited by Yahya R. Kamalipour and Hamid Mowlana, Westport, Conn.: Greenwood Press, 1994.

Kellner, Douglas. *The Persian Gulf TV War.* Boulder, Colo.: Westview Press, 1992.

Khulaifi, Muhammed S. "Tathir al-Internet fi Mujtama'a min Wajha Nathar Tolab al Marhala al-Jaamaiyyah fi Mumlika al Arabiah al-Saudia." Paper presented at the Islamic Information Resources Conference, Riyhad, Saudi Arabia, December 1998.

Kristianasen, Wendy. "Kuwait's Islamists, Officially Unofficial." *Le Monde Diplomatique*, June 4, 2002, p. 2. http://www.mondediplo.com/2002/06/04kuwait.

"Kuwait to Add Internet to Government Schools." *Kuwait Times*, October 13, 2001, p. 1.

"Kuwait Closes al-Jaseera Office." *Al-Bawaba.com*, November 4, 2002 p. 1.

Kuwait: A MEED Practical Guide. London: Middle East Economic Digest Limited, 1985.

Kuwait Times, January, 19, 1997, p. 4.

Kuwait Times, January 22, 1997, p. 20.

Kuwait Times, January 9, 2002, p. 1.

"The Kuwaiti Way of Life." Embassy of the State of Kuwait, p. 1. http://www.kuwait-embassy.org (accessed Feb. 11, 2004).

"Kuwaiti Women Strongly Resent Idea of Staying at Home." *Arab Times*, January, 8, 1994, p. 1.

"Kuwaitiun wal Kitabun!" (Kuwaitis and Books!). *Al-Rai al-Am*, August, 30, 1997, p. 13.

Landers, Jim. "Separate but Equal?" *Dallas Morning News*, March 9, 2003, p. 2J.

Larson, James F. and Heung-Soo Park. *Global Television and the Politics of the Seoul Olympics* Boulder, Colo.: Westview Press, 1993.

Levins, John. *Days of Fear: The Inside Story of the Iraqi Invasion and Occupation of Kuwait.* Dubai: Motivate Publishing, 1997.

Liff, Sonia, Fred Steward, and Peter Watts. "New Public Places for Internet Access: Networks for Practice-Based Learning and Social Inclusion." In *Virtual Society: Technology, Cyberbole, Reality,* edited by Steve Woolgar, 78–98. Oxford: Oxford University Press, 2002.

Loader, Brian D. "The Governance of Cyberspace: Politics, Technology, and Global Restructuring." In *The Governance of Cyberspace,* edited by Brian D. Loader, London: Routledge, 1997.

Longva, Anh Nga. *Walls Built on Sand: Migration, Exclusion, and Society in Kuwait.* Boulder, Colo.: Westview Press, 1997.

Los Angeles Times. November 11, 2001, p. 4.

Low, Linda. "Social and Economic Issues in an Information Society: A Southeast Asian Perspective." *Asian Journal of Communications* 6:1 (1996): 12.

Machiavelli, Niccolo. *Il Principe.* Skokie, IL: Distribooks, 1997.

Machinchick, Kelly. "Kuwaiti Women's Suffrage Movement Comes to Washington." *Washington File,* February 11, 2003, p. 1.

Marvin, Carolyn. *When Old Technologies Were New: Thinking about Electric Communication in the Late Nineteenth Century.* Oxford: Oxford University Press, 1988.

al-Mazeedi, Moosa L., and Ibrahim A. Ismail. "The Educational and Social Effects of the Internet on Kuwait University Students." Paper presented at the Conference on Information Superhighway, Safat, Kuwait, March 3–5, 1998.

McConnell International. "Ready? Net. Go! Partnerships Leading the Global Economy." http://www.mcconnellinternational.com/ereadiness/ereadinessreport2.htm (accessed Oct. 30, 2001).

McDonald, Fiona. "In for a Cuppa . . . Dot Com." *Arab Times,* March 20, pp. 1, 8.

McLuhan, Marshall. *The Gutenberg Galaxy: The Making of Typographic Man.* Toronto: University of Toronto Press, 1962.

"Media in Kuwait." *World Press Freedom Review 1999–2001.* http://www.freemedia.at/wpfr/Kuwait.htm (accessed Sept. 30, 2001).

Hafez, Kai, ed., *Mass Media, Politics and Society in the Middle East.* Cresskill, N.J.: Hampton Press, 2001.

al-Menayes, Jamal. "Television Viewing Patterns in the State of Kuwait after the Iraqi Invasion." *Gazette: A Journal of International Communications* 56 (1996):121–134.

Miller, Daniel, and Don Slater. *The Internet: An Ethnographic Approach.* Oxford: Berg, 2001.

Ministry of Planning. *Kuwait and Social Development: Leadership, Planning, Popular Participation, and Humanitarian Orientation.* Safat: Center for Research and Studies on Kuwait, 1995.

The Mosaic Group. "Global Diffusion of the Internet Project: An Initial Inductive Study." http://mosaic.unomaha.edu/GDI1998/GDI1998.html (accessed June 23, 1998).

"Movies Censored by Ministry, Says Official." *Arab Times,* March 20–21, 1997, p. 4.

al-Mughni, Haya. "From Gender Equality to Female Subjugation: The Changing Agendas of Women's Groups in Kuwait." In *Organizing Women: Formal and Informal Women's Groups in the Middle East,* edited by Dawn Chatty and Annika Rabo, 195–210, Oxford: Berg.

_____. *Women in Kuwait: The Politics of Gender.* London: Saqi Books, 1993.

Muntada al-Marah wa Sana' al-Qirar: Bahath wa-Awraq al-Amal. Kuwait: Women's Cultural and Social Society, 1996, 61–73.

al-Najjar, Ghanim. "Challenges Facing Kuwaiti Democracy." In *Middle East Journal* 54:2 (Spring, 2000): 242–58.

Naughten, John. *A Brief History of the Future: The Origins of the Internet.* Collingdale, UK: Diane Publishing, 1999.

Neustadlt, Alan, John P. Robinson, and Meyer Kestnbaum. "Doing Social Research Online." In *The Internet and Everyday Life.* Oxford: Blackwell, 2002.

Norris, Pippa. *The Digital Divide: Civic Engagement, Information Poverty, and the Internet Worldwide.* New York: Cambridge University Press, 2001.

Norton, Augustus R.. "New Media, Civic Pluralism, and the Slowly Retreating State." In *New Media in the Muslim World: The Emerging Public Sphere.* edited by Dale F. Eickelman and Jon W. Anderson, 15–30. Bloomington: Indiana University Press, 1999.

Nye, Joseph S. Jr. and William A. Owens. "America's Information Edge: The Nature of Power." *Foreign Affairs* vol. 75, #2, March/April (1996): p. 23.

Osman, Khalil. "Kuwaiti 'royal family' Trying to Mend a Bursting Dam with Sticky Tape." *Crescent International* (February 2001): 3 http://www.muslimedia.com/arcives/oaw01/kuw-burst.htm.

"Pan-Arab Mobile Phone Subscribers Reach 30 Million Mark in 2003." *AME Info,* January 27, 2004, p. 1. http://www.ameinfo.com.

PC Middle East. "Almost One Million Online in Arab Countries." August 5, 1999, p. 1. http://www.ditnet.co.ae/itnews/newsaug99/newsaug5.html.

"PC Penetration vs. Internet User Penetration in GCC Countries." *Madar Research Journal: Knowledge, Economy and Research on the Middle East* 1(October 2002): 1–15.

Pew Charitable Trust. *Internet and American Life.* http://www.pewinternet.org (accessed Dec. 5, 2002).

Plato. *The Phaedrus.* Translated by Benjamin Jowett. http://www.classics.mit.edu/Plato/phaedrus.html (accessed Feb. 22, 2003).

"Politics Takes Backseat in Kuwait Poll." *Al-Ahram Weekly*, 646 (July 10–16 2003): 1. http://www.weekly.ahram.org.eg/2003/646/re2.htm.

Poster, Mark. *What's the Matter with the Internet?* Minneapolis: University of Minnesota Press, 2001.

Postman, Neil. *Technopoly: The Surrender of Culture to Technology*. New York: Vintage, 1993.

Prusher, Ilene R. "Rise of Islamists Veils Liberalism." *Christian Science Monitor* April, 24, 2000, p. 1. http://www.csmonitor.com/durable/2000/04/24/fp8sl-csm.shtml (accessed 2000).

Putnam, Robert. *Bowling Alone*. New York: St. Martin's Press, 2000.

al-Qabas. January 17, 1997, p. 21.

al-Qabas. June 21, 1997, p. 15.

al-Qaradawi, Dr. Yusif. *Makanah al-Marah fi Islam* (The Status of Women in Islam). Cairo: Islamic Home Publishing Co., 1997

al-Qinanie, Yusif. *Pages from Kuwait's History* (Arabic). Kuwait: Government Printing House, 1968.

al-Rai al-Am. August 29, 1997, p. 12.

Rainie, Harrison "Lee." "Forward" in *Society Online: The Internet in Context*, edited by Philip N. Howard and Steve Jones, i–vi. Thousand Oaks, Calif.: Sage, 2004.

Randall, Neil. *The Soul of the Internet: Net Gods, Netizens and the Wiring of the World*. London: International Thomson Computer Press, 1996.

Rashoud, Claudia Farkas. *Kuwait: Before and After the Storm*. Kuwait City: The Kuwait Bookshop, 1992.

Reporters without Borders. "Kuwait 2003 Annual Report," p. 1. http://www.rsf.org (accessed 2003).

Rochdi, Najat. "Cultural Boundaries and Cyberspace." *Women's Wire*. http://www.womenswire.org/opinion.htm (accessed May 22, 2002).

Rochlin, Gene I. *Trapped in the Net: The Unanticipated Consequences of Computerization.* Princeton: Princeton University Press, 1997.

Roland, Wade. *Spirit of the Web: The Age of Information from the Telegraph to the Internet.* Toronto: Key Porter Books, 1999.

Rumaihi, Muhammad. *Beyond Oil: Unity and Development in the Gulf.* London: Al-Saqi Books, 1986.

_____. *al-Kuwait lyisa al-Naftan,* Beirut: Dar al-Jadid, 1995.

Salus, Peter H. *Casting the Net: From ARPANET to Internet and Beyond.* New York: Addison-Wesley, 1995.

Sayyid, Bobby S. *A Fundamental Fear: Eurocentrism and the Emergence of Islamism.* London: Zed Books, 1997.

Seattle Times. July 1, 2002, p. A3.

al-Sharq al-Awsat. October 13, 2001, p. 1.

Smith, Simon C. *Kuwait, 1950–1965: Britain, the al-Sabah, and Oil.* Oxford: Oxford University Press, 1999.

Spender, Dale. *Nattering on the Net.* Toronto: Garamond Press, 1995.

Standage, Tom. *The Victorian Internet: The Remarkable Story of the Telegraph and the 19th Century's On-Line Pioneers.* New York: Berkley Publications Group, 1999.

The Star. July 23, 1998. http://www.star.arabia.com/980730/TE2.html.

Stowe, Harriet Beecher. *Pink and White Tyranny.* New York: Macmillan, 1989.

Strobel, Warren P. *Late-Breaking Foreign Policy: The News Media's Influence on Peace Operations.* Washington, D.C.: U.S. Institute of Peace, 1997.

Tacchi, Jo, Don Slater, & Peter Lewis. "Ethnographic Action Research Approach." Paper presented at the IT4D Conference, Wolfson College, Oxford University, July 18, 2004.

"Tackle Effects of Media on Islamic National Values." *Arab Times,* March, 10, 1997, p. 2.

Talbot, Strobe. "Democracy and the National Interest." *Foreign Affairs* 75:6 (November/December, 1996): 47–63.

Talero, Edward, and Phillip Gaudette. "Harnessing Information for Development: A Proposal for a World Bank Group Strategy." *World Bank* (March 1995): p. 2. http://www.worldbank.org/html/fpd/harnessing.

Taylor, Philip M. *War and the Media: Propaganda and Persuasion in the Gulf War.* 2d ed. Manchester: University of Manchester Press, 1998.

"Terrorists On-Line." *Arab Times,* January, 26, 1997, p. 1.

Tétreault, Mary Ann. "Civil Society in Kuwait: Protected Spaces and Women's Rights." *Middle East Journal,* 47 (Spring 1993): 275–91.

_____. "Frankenstein's Lament in Kuwait." *Foreign Policy in Focus,* November, 29, 2001, p. 3. (www.fpif.org/commentary/2001/0111kuwait.body.html).

_____. *Stories of Democracy: Politics and Society in Contemporary Kuwait.* New York: Columbia University Press, 2000.

Toumi, Ilkka. *Networks of Innovation: Change and Meaning in the age of the Internet.* Oxford: Oxford University Press, 2002.

Turkle, Sherry. *Life on the Screen: Identity in the Age of the Internet.* New York: Simon and Schuster, 1995.

Turoff, Murray, and Starr Roxanne Hiltz. "Meeting through Your Computer." *IEEE Spectrum* 14 (1997): 58–64.

UCLA Center for Communications Policy. "Surveying the Digital Future." http://www.ccp.ucla.edu (accessed Sept. 15, 2003).

UNESCO Yearbook. New York: UNESCO, 1996.

Vig, Norman J. "Technology, Philosophy and the State." In *Technology and Politics,* edited by Michael Craft and Norman J. Vig, 8–32. Durham, N.C.: Duke University Press, 1988.

Warschauer, Mark. *Technology and Social Inclusion: Rethinking the Digital Divide.* Cambridge: MIT Press, 2003.

Al-Watan. January 29, 1997. p. 1.

Al-Watan. September 17, 1997, p. 7.

Webster, Frank, ed. "A New Politics?" In *Culture and Politics of the Information Age.* London: Routledge, 2001.

Wellman, Barry, and Caroline Haythornthwaite, eds. *The Internet and Everyday Life.* Oxford: Blackwell, 2002.

West, John. "Kuwait Bikers Seek Thrills Despite Spills." *Reuters* August 24, 2003, p. 1. http://www.aduni.org/~johnwest.

Wheeler, Deborah. "The Internet and Public Culture in Kuwait." *Gazette 63 (2001): 187–201.*

_____. "New Media, Globalization and Kuwaiti National Identity." Middle East Journal 54:3 (2000): 432–44.

_____. "Living At E.Speed: A Look at Egypt's E.Readiness." In *Challenges and Reforms of Economic Regulation in the MENA Countries,* edited by Imed Limam, 129–157. Cairo: American University in Cairo Press, 2003.

_____. "Blessings and Curses: Women and the Internet in the Arab World." In *Women and the Media in the Middle East,* edited by Naomi Sakr, 138–168. London: IB Tauris, 2004.

Whitaker. Reginald. *The End of Privacy: How Total Surveillance is Becoming a Reality.* New York: New Press, 1999.

Winner, Langdon. "Do Artifacts Have Politics?" In *Technology and Politics,* edited by Michael E. Kraft and Norman J. Vig, 33–53. Durham, N.C.: Duke University Press, 1988

_____. *The Whale and the Reactor: A Search for Limits in an Age of High Technology.* Chicago: University of Chicago Press, 1986.

Winston, Brian. *Media, Technology and Society: A History, from the Telegraph to the Internet.* London: Routledge, 1998.

Woolgar, Steve, ed. *Virtual Society: Technology, Cyberbole, Reality.* Oxford: Oxford University Press, 2002.

World Economic Forum. *Arab World Competitiveness Report 2002–2003.* Oxford: Oxford University Press, 2003.

Wyatt, Sally, Graham Thomas, and Tiziana Terranova. "They

Came, They Surfed, They Went Back to the Beach: Conceptualizing Use and Non-Use of the Internet." In *Virtual Society: Technology, Cyberbole, Reality,* edited by Steve Woolgar, 23–40. Oxford: Oxford University Press, 2002.

"Young Children Find Pornography on the Net." http://www.nua.ie/surveys (accessed April 27, 2003).

Yue, Chia Siow, and Jamus Jerome Lim, eds. *Information Technology in Asia: New Development Paradigms.* Singapore: Institute of Southeast Asian Studies, 2002.

Index